RESPACE – Key Technologies for Reusable Space Systems

Results of a Virtual Institute Programme of the German Helmholtz-Association, 2003–2007

Ali Gülhan
(Editor)

Springer

Dr. Ali Gülhan
German Aerospace Center
Windtunnel Department
Linder Höhe
55147 Köln
Germany
E-mail: ali.guelhan@dlr.de

ISBN 978-3-642-09661-7 e-ISBN 978-3-540-77819-6

DOI 10.1007/978-3-540-77819-6

Notes on Numerical Fluid Mechanics
and Multidisciplinary Design ISSN 1612-2909

Printed in acid-free paper

5 4 3 2 1 0

springer.com

NNFM Editor Addresses

Foreword

A few years ago the Helmholtz Association (HGF) consisting of 15 research Institutions including the German Aerospace Center (DLR) started a network research program called 'Virtual Institutes'. The basic idea of this program was to establish research groups formed by Helmholtz research centers and universities to study and develop methods or technologies for future applications and educate young scientists. It should also enable and encourage the partners of this Virtual Institute after 3 years funding to continue their cooperation in other programs. Following this HGF request and chance the DLR Windtunnel Department of the Institute of Aerodynamics and Flow Technology took the initiative and established a network with other DLR institutes and German universities RWTH Aachen, University of Stuttgart and Technical University Munich.

The main goal of this network was to share the experience in system analysis, aerodynamics and material science for aerospace for improving the understanding and applicability of some key technologies for future reusable space transportation systems. Therefore, the virtual institute was named RESPACE (Key Technologies for Re-Usable Space Systems).

As leading concepts the Liquid Fly Back Booster (LFBB was chosen. The technological topics dealing with active cooling of hot structures and base flow phenomena of booster configurations were accomplished by combined application of experimental and numerical tools. A multidisciplinary approach was required for these tasks, which needs basic work to improve the physical modelling of numerical codes and applied research for the design and manufacturing of ground testing models for the qualification of above mentioned key technologies.

The virtual institute RESPACE improved the cooperation between the partner universities and DLR significantly, which is confirmed by the actual collaborations of the partners in the frame of DFG and ESA programmes. The financial support of Helmholtz Association (HGF) and programmatic support of German Aerospace Center (DLR) was essential for the establishment and success of this research group. The effort of Mrs. Chr. Adams in preparation of the manuscript is highly acknowledged.

The editor is grateful to Prof. E.H. Hirschel as the General Editor of the 'Notes on Numerical Fluid Mechanics and Multidisciplinary Design' and to the Springer-Verlag for the opportunity to publish the results of the RESPACE.

Köln, 08.10.2007 Ali Gülhan

Table of Contents

General Introduction

A. Gülhan

Deutsches Zentrum für Luft- und Raumfahrt (DLR)
Abteilung Windkanäle
des Instituts für Aerodynamik und Strömungstechnik
Linder Höhe, 51147 Köln, Germany
ali.guelhan@dlr.de

1 General Introduction

A few years ago the Helmholtz Association (HGF) consisting of 15 research Institutions including the German Aerospace Center (DLR) started a network research program called 'Virtual Institutes'. The basic idea of this program was to establish research groups formed by Helmholtz research centers and universities to study and develop methods or technologies for future applications and educate young scientists. It should also enable and encourage the partners of this Virtual Institute after 3 years funding to continue their cooperation in other programs. Following this HGF request and chance the DLR Windtunnel Department of the Institute of Aerodynamics and Flow Technology took the initiative and established a network with the following partners:

- DLR Windtunnel Department of the Institute of Aerodynamics and Flow Technology in Cologne (Dr. Ali Gülhan)
- DLR Space Launcher Systems Analysis Department of the Institute of Space Propulsion (Dr. Martin Sippel)
- DLR Space Structural Components Department of the Institute of Structures and Design (Dr. Hermann Hald)
- Institute of Aerodynamics, RWTH Aachen (Prof. Wolfgang Schröder)
- Shock Wave Laboratory, RWTH Aachen (Prof. Herbert Olivier)
- Transition Group of the Institute of Aerodynamics and Gasdynamics of the University of Stuttgart (Dr. Markus Kloker)
- Turbulence Simulation Group of the Institute of Aerodynamics of the TU Munich (Prof. Rainer Friedrich)

In order to perform the cooperation efficiently, the research work has been carried out in a project form and with a reference concept. The main goal was to share the experience in system analysis, aerodynamics and material science for aerospace for improving the understanding and applicability of some key technologies for future reusable space transportation systems. Therefore, the virtual institute was named RESPACE (Key Technologies for Re-Usable Space Systems).

The feasibility of future space transportation systems with respect to technical and economical aspects strongly depends on the decrease of the safety margins, i.e. mass reduction, manufacturing and maintenance costs. In addition, the reliability of space vehicles has to be improved. This challenging goal can only be achieved by a realistic

A. Gülhan (Ed.): RESPACE – Key Tech. for Reusable Space Sys., NNFM 98, pp. 1–2, 2008.
springerlink.com

system study accompanied by developments of key technologies using reliable tools with a multidisciplinary approach. Since the duration of the project was limited to three years, the activities focused on three areas:

- System study to define the technology requirements of the reference concepts,
- active cooling technologies for space transportation systems,
- base/nozzle flow phenomena of a reusable booster configuration.

To achieve these goals first the improvement of numerical and experimental tools was necessary. As leading concepts the Liquid Fly Back Booster (LFBB), which has been one of the basic concepts of the German National Space Transportation Technology Development Program ASTRA, and the new SpaceLiner vehicle defined by the DLR department for Space Launcher Systems Analysis were chosen.

For the two technological topics dealing with active cooling of hot structures and base flow phenomena of booster configurations data were accomplished by combined application of experimental and numerical tools. A multidisciplinary approach was required for these tasks, which needs basic work to improve the physical modelling of numerical codes and applied research for the design and manufacturing of ground testing models for the qualification of above mentioned key technologies.

System Requirements on Investigation of Base Flow/Plume Interaction

M. Sippel and A. Herbertz

Deutsches Zentrum für Luft- und Raumfahrt (DLR)
Systemanalyse Raumtransport (SART)
Linder Höhe, D-51147 Köln, Germany
martin.sippel@dlr.de

Summary

In order to achieve a better understanding of launcher base flow phenomena like base aerothermal heating caused by the nozzle plume external flow interaction at flight relevant conditions, the Liquid Fly Back Booster (LFBB) configuration has been used as reference. The under expanded flow condition of the Vulcain 3 engine at an altitude of about 50 km has been considered as baseline of the nozzle design. Since the flight parameter cannot be duplicated in the windtunnel H2K and a parametric study on the influence of all main parameters was necessary, nozzles with different expansion ratios have been specified and operated at different operation conditions during windtunnel tests.

1 Introduction

Rocket engine nozzles typically do not have an adaptable expansion ratio, i.e. they have a fixed nozzle exit pressure. Because of the ever decreasing surrounding pressure during ascent this leads to underexpansion and as a result widely diverging exhausts plume (Fig. 1). For launcher designs with multiple engines per stage it is necessary to understand the accompanying flow phenomena. Within the Respace program more understanding of these phenomena is created by doing windtunnel tests in combination with numerical simulations.

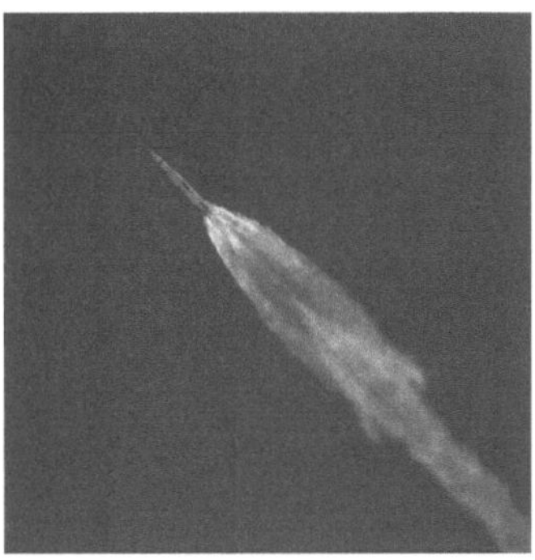

Fig. 1. Diverging exhaust plume of the Saturn V Launcher

A. Gülhan (Ed.): RESPACE – Key Tech. for Reusable Space Sys., NNFM 98, pp. 3–19, 2008.
springerlink.com

2 Determination of Boundary Conditions

2.1 Surrounding Pressure during Ascent

Representative surrounding pressure during ascent is based on the simulation of the ascent trajectory of the ASSC-2 ASTRA concept. These data were used as the

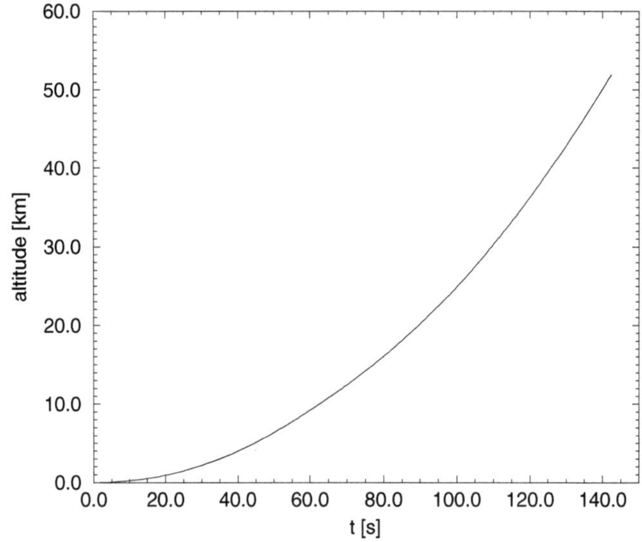

Fig. 2. Altitude as a function of time for LFBB variant Y-8

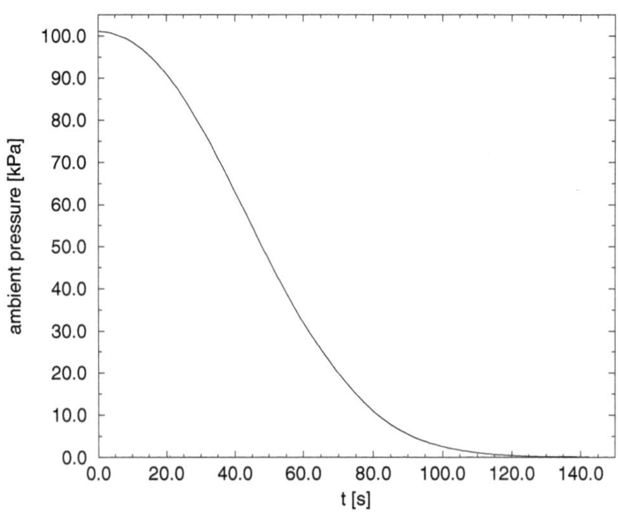

Fig. 3. Surrounding pressure as a function of time for LFBB variant Y-8

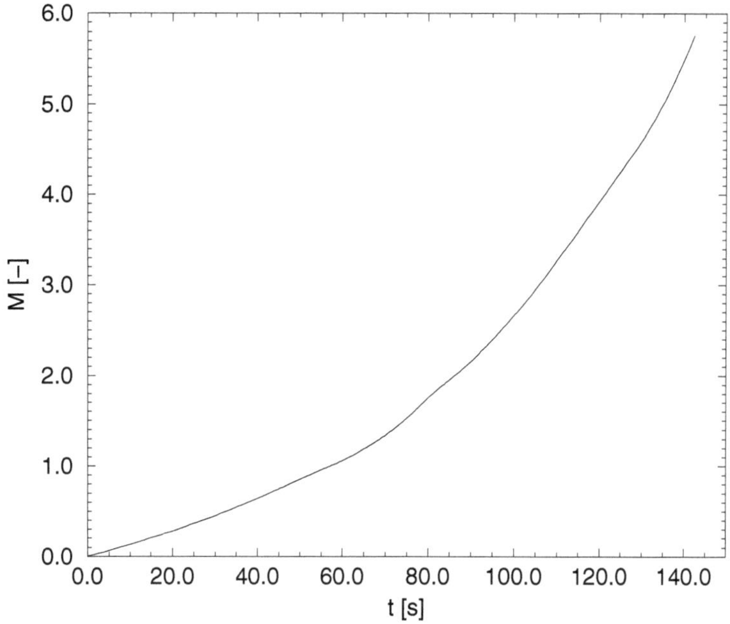

Fig. 4. F Mach number as a function of time for LFBB variant Y-8

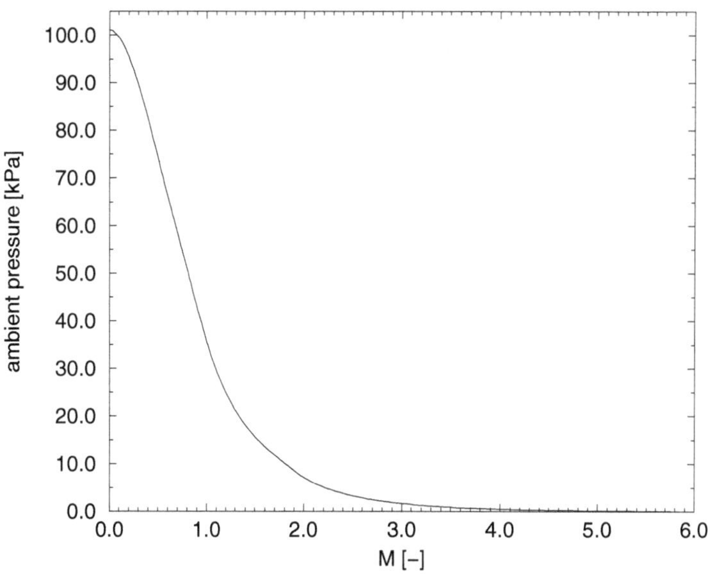

Fig. 5. Surrounding pressure as a function of time for LFBB variant Y-8

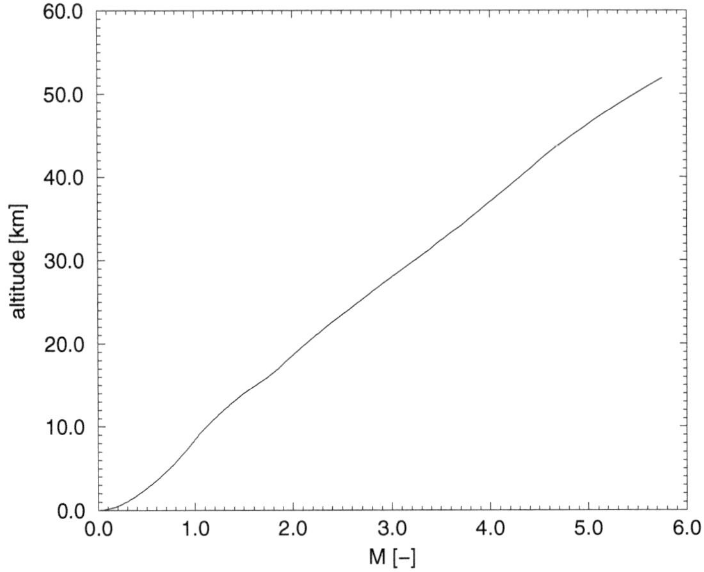

Fig. 6. Altitude as a function of Mach number for LFBB variant Y-8

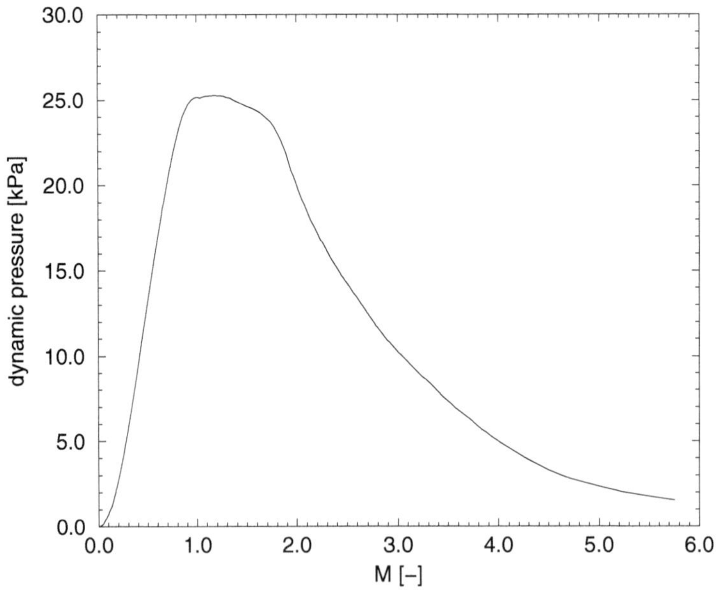

Fig. 7. Stagnation pressure as a function of Mach number for LFBB variant Y-8

boundary conditions for the base flow research. Figure 2 to 7 show representative data for the ascent trajectory of the LFBB variant Y-8. For tests in the hypersonic windtunnel H2K at DLR-Cologne a fixed Mach number of 5.3 is chosen due to technical limitations. This is in within the range of the interesting flow region with low surrounding pressure. At a flight Mach number of 5.3 at time $t = 138.2$ s and an altitude of almost 50 km, the surrounding pressure is about $p_a = 99.5$ Pa. A little later booster separation takes place at $t = 142.4$ s, $M = 5.75$ and $p_a = 66.7$ Pa. Stagnation pressure is about 2 kPa.

2.2 Flow Exit Conditions and Geometrical Boundary Conditions for the Rocket Engine

The engine which is foreseen for the LFBB is a non existing gas generator engine with an expansion ratio of $\varepsilon = 35$ [10] derived from the European Vulcain family. The engine will be called Vulcain 3-35 from now on. The assumed engine parameters are representative for big LOX/LH2 booster engines. The Vulcain 3 engine was modeled using the program LRP [3].

The program NCC [4] is used to define the engine contour. Table 2 shows the results. The definition of the parameters can be found in Fig. 17.

Table 1. Flow exit conditions of the Vulcain 3 nozzle

Exit pressure	39920 Pa
Exit temperature	1781,1 K
Exit velocity	4046,4 m/s
Exit Mach number	3,89

Table 2. Calculated contour data of the Vulcain 3 nozzle

Vulcain 3 (ε=35)		
Propellant combination:	LOX/LH2	
Char. chamber length:	1	m
Chambertemperature:	3615.3	K
Chamberpressure:	13.93	MPa
Massflowrate:	373.6	kg/s
Vacuum Isp:	413.9	s
Sealevel Isp:	362.1	s
Vacuum thrust:	1516.4	kN
Sealevel thrust:	1326.7	kN
Total length:	2.283	m
inner contour radius:	2.605	–
upstream throat radius:	1.1073	–
downstream throat radius:	0.4981	–
upstream contour angle:	25	deg
Throat velocity:	1437.3	m/s
Chamber volume:	0.05887	m^3

Table 2. *(continued)*

subsonic part length:	0.478	m
cylindric part length:	0.195	m
Throat radius:	0.137	m
Throat area:	0.05887	m²
Contraction ratio:	2.5	–
Exit diameter:	1.62	m
supersonic part length:	1.805	m
Nozzle entry angle:	33.5	deg
Nozzle exit angle:	10.1	deg
Divergence factor:	0.992	–
Expansion ratio:	35	–
nozzle exit pressure:	0.04	MPa
nozzle exit Mach number:	3.89	–
Conv. (chamber) surface:	0.5883	m²
Div. (nozzle) surface:	6.116	m²
Total surface:	6.704	m²

3 Choosing the Nozzle Expansion Ratio

Exact modeling of the boundary conditions given in chapter chapter 2.2 is not possible in windtunnel tests. Therefore, certain flow characteristics and ratio's should be chosen and only these should be achieved during windtunnel tests. The flow Mach number in the windtunnel can be set to exactly the 'real' value, but static pressure in the H2K has an upper limit of 400 Pa. The lower limit lies at 100 Pa, in case the tunnel is evacuated and no flow is present.

Total conditions in the nozzle during tests will deviate from the real nozzle even more. The model uses pre-pressurized and pre-heated air. Chamber pressure and temperature therefore cannot reach the same conditions as in the real engine. By variation of nozzle parameters, the model nozzle which simulates the real conditions the best, can be chosen.

For determination of thermodynamic properties in the model nozzle and at the model nozzle exit, CEA [2] was used. Calculations were done while varying different parameters. The following figures are the basis for choosing the model nozzle expansion ratio. Variations are limited to conditions obtainable in the windtunnel.

The air is assumed to be of the following composition

$$\psi_{N2} = 78.0881\%$$

$$\psi_{O2} = 20.9495\%$$

$$\psi_{Ar} = 0.9324\%$$

$$\psi_{CO2} = 0.03\%$$

All calculations take place at chemical equilibrium.

Figure 8 and Fig. 9 show that the 'real' exit pressure of 40 kPa can be achieved in different combinations.

Figure 10 shows that the ‚real' exit Mach number of 3.88 is achieved with an expansion ratio of 9.

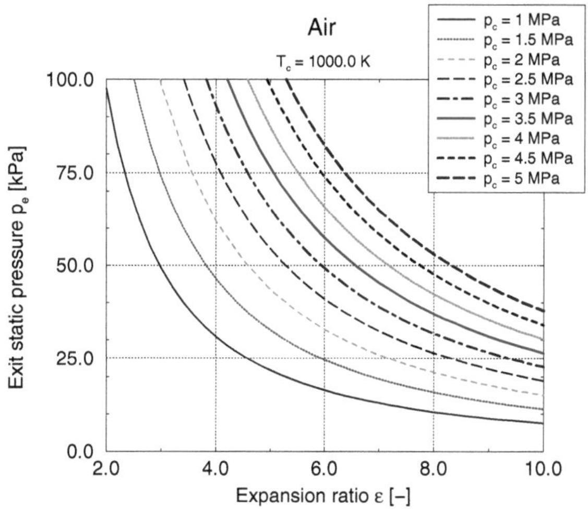

Fig. 8. Static pressure at nozzle exit for different pressures vs. expansion ratio ($2 \leq \varepsilon \leq 5$)

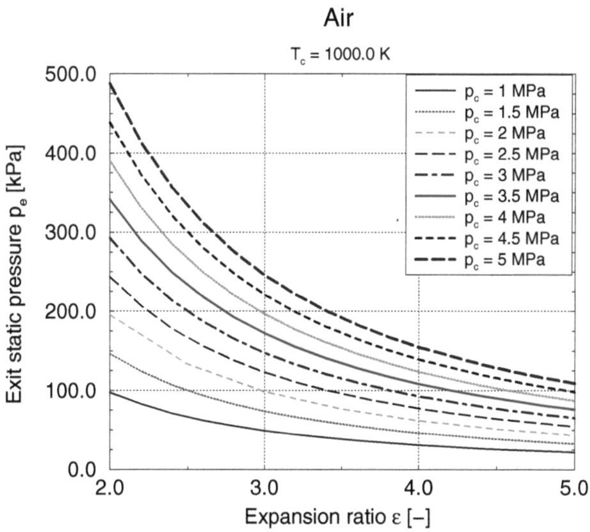

Fig. 9. Static pressure at nozzle exit for different total pressures vs. expansion ratio ($2 \leq \varepsilon \leq 10$)

Figure 11 shows that the total temperatures obtainable in the windtunnel already are less than the 'real' nozzle exit temperature; therefore this nozzle exit temperature can never be achieved.

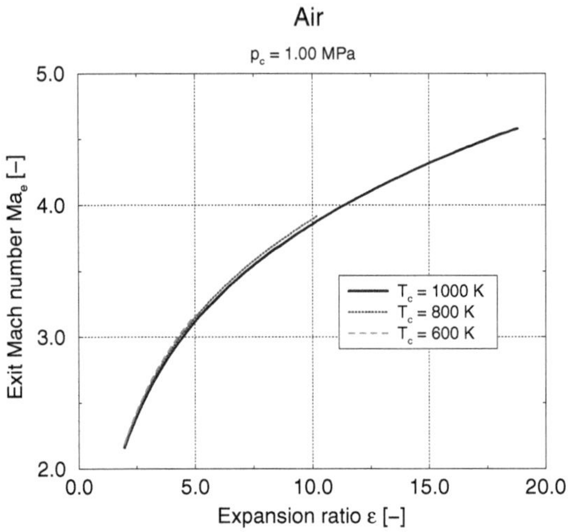

Fig. 10. Mach number at nozzle exit for different total temperatures vs. expansion ratio

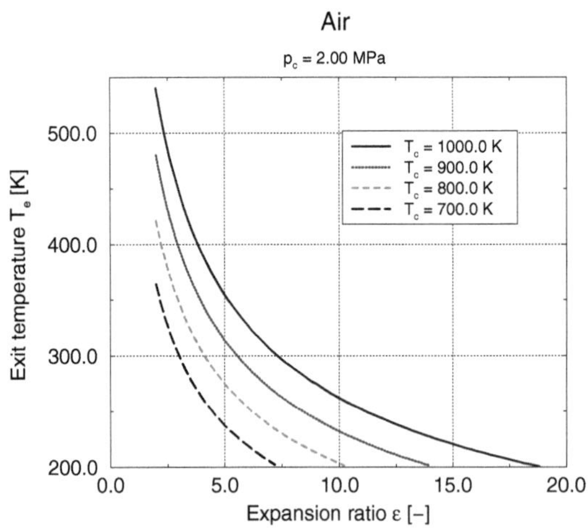

Fig. 11. Static temperature at nozzle exit for different total temperatures vs. expansion ratio

Figure 14 shows that for these total temperatures, the 'real' exit velocity can never be obtained either.

Mass flow through the nozzle is about 400-500 g/s. For this mass flow, total conditions for the flow through the model nozzle are set to a pressure $p_c = 2$ MPa and a temperature of 900 K. Because of the windtunnel limitations two model nozzle exit ratio's are chosen. The higher exit ratio ($\varepsilon = 10$) simulates the Mach number of the reference engine. The lower expansion ratio ($\varepsilon = 5$), simulates the pressure ratio's between nozzle exit und surroundings.

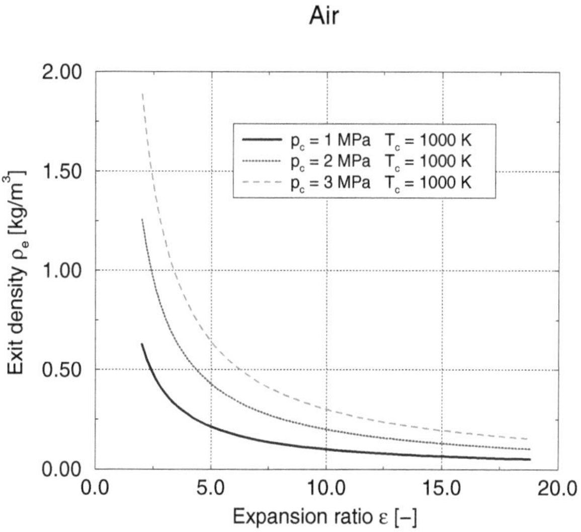

Fig. 12. Density at nozzle exit for different total pressures vs. expansion ratio

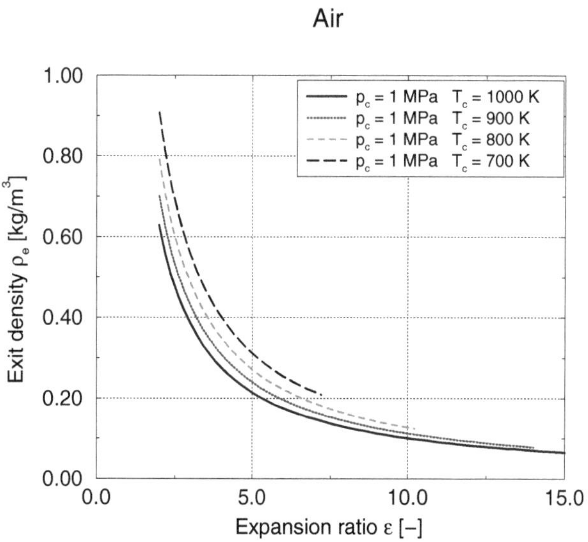

Fig. 13. Density at nozzle exit for different total temperatures vs. expansion ratio

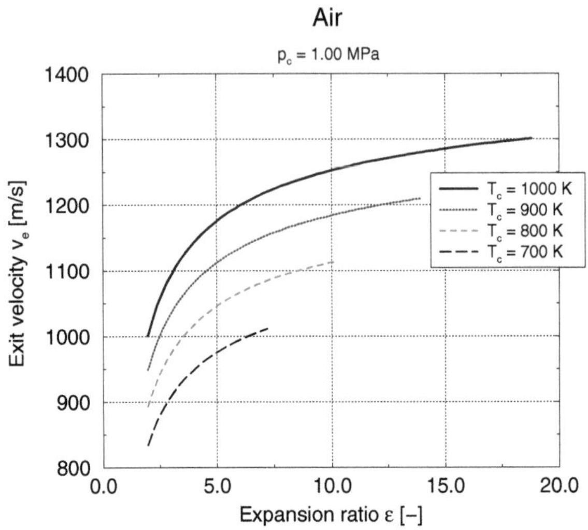

Fig. 14. Velocity at nozzle exit for different total temperatures vs. expansion ratio

4 Design of the Model Nozzles

4.1 Simplifications Used during the Contour Design

For the design of the model nozzles the program NCC is used [4]. The supersonic nozzle contour is determined by flow calculations using the method of characteristics. The generated contour depends on the specific heat coefficient κ. Its value changes along the nozzle axis as a function of temperature and gas composition. NCC uses the

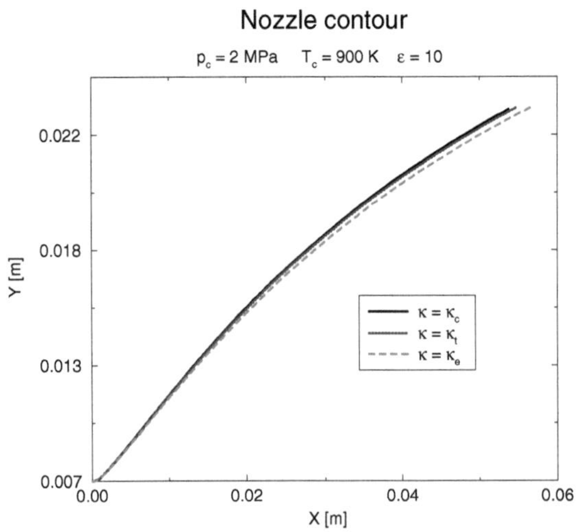

Fig. 15. Contours for different κ

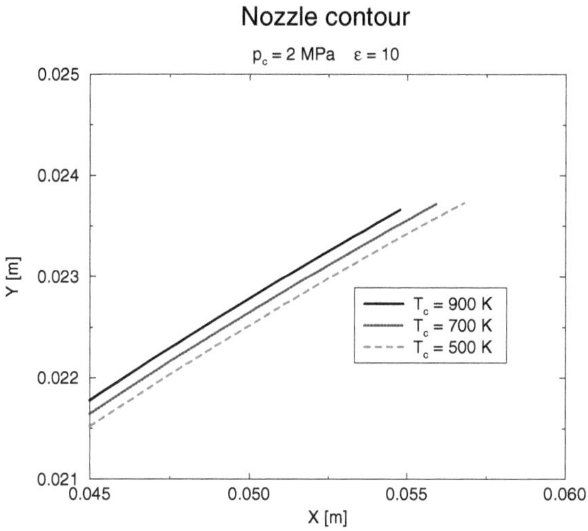

Fig. 16. Nozzle exit contours for different temperatures

approximation of a constant. Figure 15 shows a sensitivity analysis of the contour for different κ. Values for κ_c, κ_t and κ_e (combustion chamber, throat and nozzle exit) are generated by CEA. For the design of the nozzle the value at the throat (κ_t) is chosen.

As a result of the dependence of κ on temperature, the contour shows a dependence on total temperature. Figure 16 shows this dependency. At the nozzle exit a difference of up to 1mm can be observed.

4.2 Definition of Nozzle Contours

Figure 17 shows the design variables used by NCC of the nozzle contour. Table 3 shows the value of these variables for both model nozzles and the real nozzle.

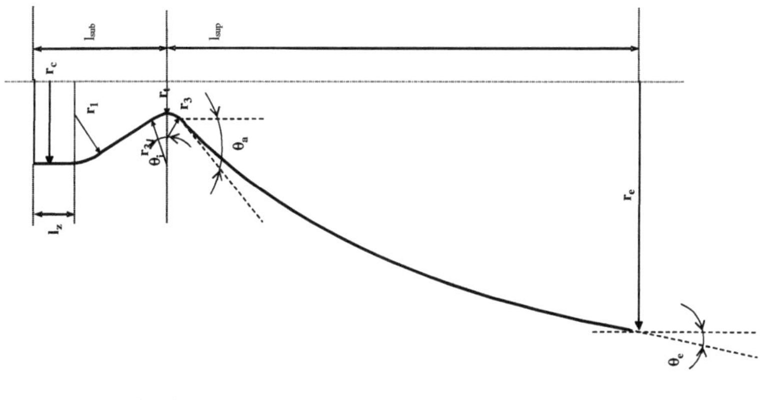

Fig. 17. Definition of relevant thrust chamber variables

Table 3. Values of the thrust chamber variables for the two model nozzlea and the real nozzle (definition according to Fig. 17)

		nozzle 1 ($\varepsilon = 5$)	nozzle 2 ($\varepsilon = 10$)	real nozzle ($\varepsilon = 35$)
Chamber radius	r_c	30 mm	30 mm	216,5 mm
Throat radius	r_t	7,5 mm	7,5 mm	136,9 mm
Exit radius	r_e	16,73 mm	23,66 mm	809,9 mm
Radius of curvature at end of cylinder	r_1	0 mm	0 mm	356,6 mm
Radius of curvature downstream of throat	r_2	8,3 mm	8,3 mm	151,6 mm
Downstream contour angle	θ_i	25°	25°	25°
Radius of curvature downstream of throat	r_3	3,74 mm	3,74 mm	68,2 mm
Nozzle inlet angle	θ_a	13,41°	16,89°	33,5°
Nozzle exit angle	θ_e	10,24°	9,98°	10°
Length of cylinder	l_z	101 mm	101 mm	195 mm
Length of subsonic part (incl. cyl.)	l_{sub}	134 mm	134 mm	478 mm
Length of supersonic part	l_{sup}	34 mm	54,8 mm	1805 mm

Fig. 18. CAD picture of the model nozzle ($\varepsilon = 10$) in I-DEAS

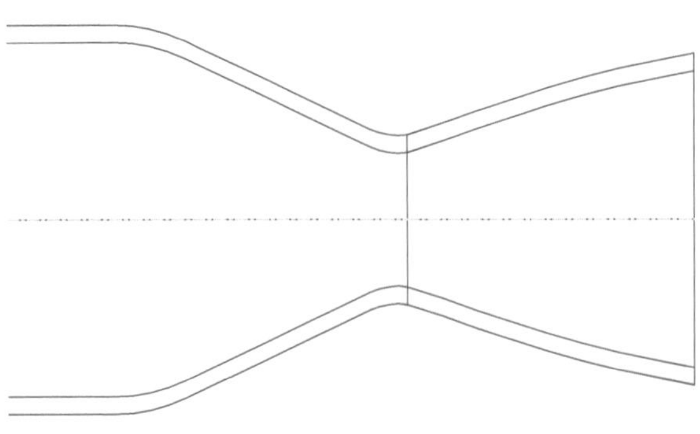

Fig. 19. Cross section view of the smaller model nozzle ($\varepsilon = 5$)

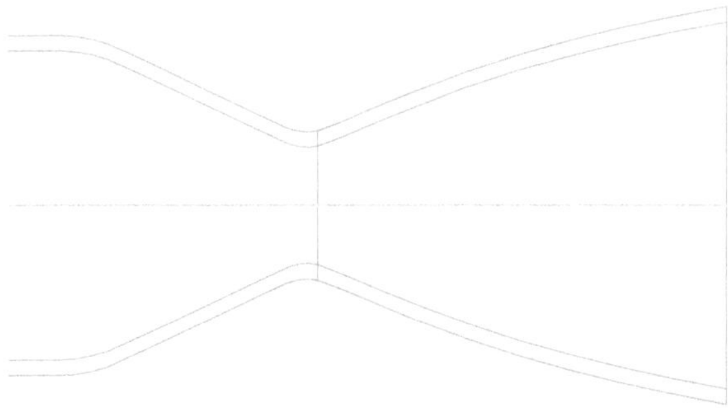

Fig. 20. Cross section view of the bigger model nozzle (ε = 10)

Figure 18 shows a CAD picture of the NCC generated model nozzle with highest expansion ratio. The long cylindrical part makes sure the heated air reaches laminar conditions.

Figure 19 shows a cross section of the NCC generated smaller nozzle. Note that combustion chamber is only shown partially. Figure 20 shows the bigger nozzle.

4.3 Expected Flow Conditions at the Nozzle Exit

Expected flow conditions at the nozzle exit are calculated using one and two dimensional analyses. During testing the combustion chamber pressure ideally is 2 MPa and the total temperature is 900 K. Because the pressure losses are not known, additional analysis is done for a total pressure of 1.8 MPa. Exit conditions are listed in Table 4. These values are calculated using CEA [2]. All values are calculated assuming a total temperature of 900 K. Π_{vac} represents the pressure ratio between nozzle exit pressure and the surrounding pressure with an evacuated windtunnel. In this case, surrounding pressure p_a is 100 Pa. In operation ($M = 5.3$), $p_a = 400$ Pa.

The Reynolds number is defined using the nozzle exit diameter

$$Re = \frac{d_e \, \rho_e \, v_e}{\eta_e}$$

Table 4. Expected flow conditions at the nozzle exit of the model nozzles

		p_e	M_e	Π_{vac} *	Π **	$\dot{m}$	Re
		Pa	–	–	–	kg/s	–
Nozzle 1 (ε = 5)	p_c = 2 MPa	43100	3.13	431	107.8	0.47	929145
Nozzle 1 (ε = 5)	p_c = 1,8 MPa	38790	3.13	387.9	97	0.42	836230
Nozzle 2 (ε = 10)	p_c = 2 MPa	14970	3.88	149.7	37.4	0.47	836566
Nozzle 2 (ε = 10)	p_c = 1,8 MPa	13470	3.88	134.7	33.7	0.42	752887

* at $M = 0$ ** at $M = 5.3$

For the 'real', reference case, flow conditions are listed in Table 5. It can be seen that for the model nozzle with expansion ratio 5, the exit pressure is close to the reference case. In case of the evacuated windtunnel, also pressure ratio between nozzle exit and surrounding pressure can be simulated very well. However, neither the correct nozzle Mach number nor interaction with surrounding flow is achieved. In case of the small nozzle in the Mach 5.3 flow it is still possible to achieve pressure ratio's of over 100. This strongly exceeds the research results of NASA [8, 11]. It can be expected that strong divergence in the exit flow will be present. Nozzle 2 achieves complete agreement in the exit Mach number. However, the pressure ratio is limited because of the smaller exit pressure. On the other hand, the ratio between nozzle exit diameter and base diameter of the vehicle is more in agreement with the reference case. Therefore the investigation of the interaction between surrounding flow and nozzle exit flow can produce more realistic results.

Table 5. Flow conditions of the reference case

Flight Mach number	M	5.3
Reynolds number of surrounding flow (based on d_e)	Re	211568
Surrounding pressure	p_a	99.5 Pa
Mach number at nozzle exit	M_e	3.89
Reynolds number of nozzle flow (based on d_e)	Re	4659073
Nozzle exit pressure	p_e	39920Pa
Pressure ratio	Π	401

Reynolds numbers for the tests are about 5 times lower as for the real conditions. However, results of previous research indicate that for base flow with nozzle exit flow, viscosity effects are of minor importance [8, 11].

Previous data are determined by a 1 dimensional flow analysis. With the NASA program TDK [1] a 2 dimensional analysis is made of the nozzle with expansion ratio of 10. Compressibility is taken into account and a turbulent boundary layer is assumed. The assumption on the boundary layer is not verified, but for the two dimensional analysis the influence is of little importance. A further assumption is an adiabatic wall. This is justified by the fact that the material thickness of the model nozzle is relatively high, the test times are short and the flow is only heated fairly.

TDK calculates radial pressure and velocity distribution of the nozzle. Total pressures of 1 and 2 MPa are used. Figures 21 and 22 respectively show the velocity and pressure distributions in the exit surface. For the velocity the following holds

$$v_e = \sqrt{v_x^2 + v_y^2}$$

From center to edge (no boundary layer), the absolute velocity decreases with about 4.5% (55 m/s). Velocities in the case of the 2 MPa chamber pressure and the 1.8 MPa chamber pressure hardly differ. The exit pressure shows a small difference of about 1.5 kPa Fig. 22.

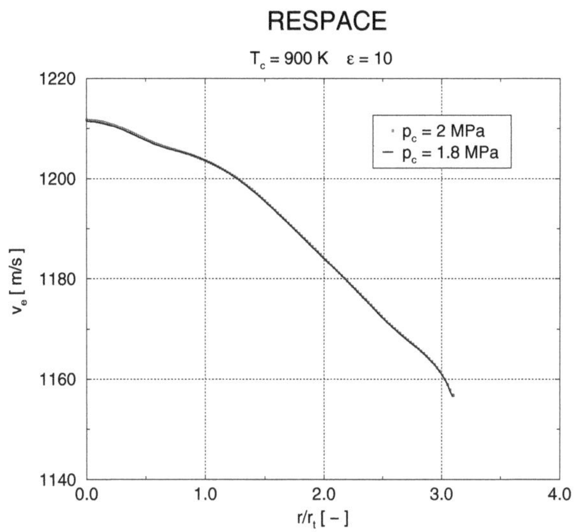

Fig. 21. Exit velocity according to TDK in radial direction normalized over the throat radius

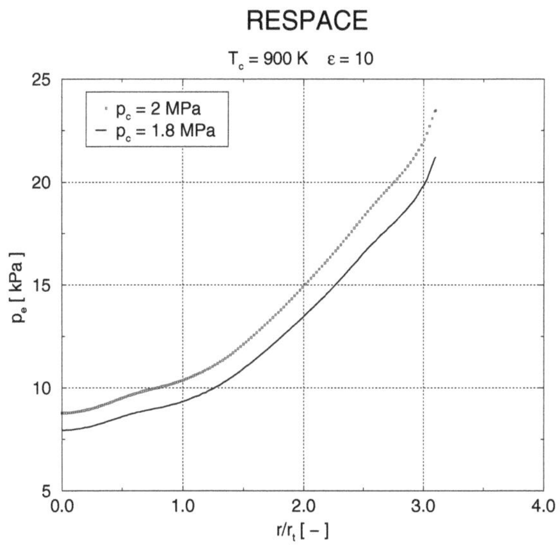

Fig. 22. Static exit pressure according to TDK in radial direction normalized over the throat radius

Figure 23 shows that qualitative pressure ratio calculated by the program is independent of chamber pressure. The figure shows the pressure distributions for different chamber pressures normalized over the axial exit pressure.

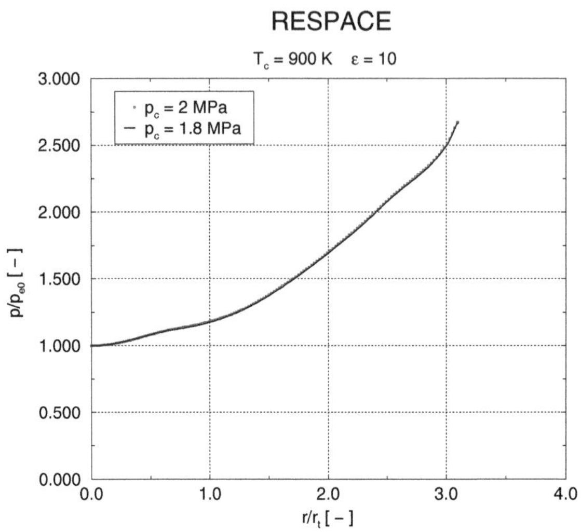

Fig. 23. Normalized exit pressure according to TDK in radial direction normalized over the throat radius

Some of the most important flow parameters determined by using TDK are listed in Table 6. According to this, the thickness of the boundary layer at the exit is about 0.5 mm. The thrust of the model nozzle with $\varepsilon = 10$ is expected to be about 570 N.

Table 6. Expected flow Ccnditions at nozzle exit of the model nozzle according to two dimensional analysis with ε = 10 and combustion chamber pressure of 1.8 and 2.0 Mpa

Total pressure	[Mpa]	1.800	2.000
Total temperature	[K]	900	900
Boundar layer parameters			
δ (exit)	[cm]	5.26E-02	5.12E-02
δ (throat)	[cm]	2.03E-03	2.00E-03
Thrust chamber data			
Thrust	[kN]	0.5121	0.5691
c^* (TDK)	[m/s]	761.69	761.6
$\dot{m}$ (TC)	[kg/s]	0.4168	0.4632
Isp (TC)	[Ns/kg]	1228.561	1228.686

References

1. Dunn, S., et al.: Two-Dimensional Kinetics (TDK 91/PRO) Nozzle Performance Computer Program. In: Engineering Methods. USA, vol. I (1991)
2. Gordon, S., McBride, B.J.: Computer Program for Calculations of Complex Chemical Equilibrium Compositions and Applications. In: Users Manual and Program Description. NASA Lewis Research Center, NASA RP-1311, vol. II (1996)

3. Herbertz, A.: Guidelines for Rocket Engine Cycle Visualisation, using LRP, Excel® and Visio®. Deutsches Zentrum für Luft- und Raumfahrt e.V., Köln, SART TN-005/2002 (2003)
4. Herbertz, A., Défosse, X., Sippel, M.: Beschreibung des Düsenkontur-Auslegungsprogramms NCC. Deutsches Zentrum für Luft- und Raumfahrt e.V., Köln, DLR-IB 645-2001/12 (2001)
5. Herbertz, A., Sippel, M.: Vorbereitung der Untersuchungen zur Heckströmung und Plume-Entwicklung bei Raumtransportsystemen. DLR-IB 647-2004/04, SART TN-005/2004 (2005)
6. Kleber, A.: Auslegung von unkonventionellen Raketendüsen. Deutsches Zentrum für Luft- und Raumfahrt e.V., Lampoldshausen, DLR-IB 645-97/9 (1997)
7. Kloft, S.: Auslegung eines Windkanalmodells zur Untersuchung der Interaktion zwischen unterexpandierter Düsenströmung und Raketenumströmung. Diplomarbeit, Fachhochschule Koblenz (2004)
8. Love, E.S., et al.: Experimental and Theoretical Studies of Axisymmetric Free Jets. NASA Langley Research Center, NASA TR R-6 (1959)
9. Sippel, M., et al.: Critical Aerothermal Phenomena on a Reusable Booster Stage. In: 4th Intern. Symp. Atmospheric Reentry Vehicles & Systems (2005)
10. Sippel, M., et al.: Documentation on the Design and Analyses of the Liquid Fly Back Booster Concept ASSC-2 Y. ASTRA Doc. No. 1000-012, Issue: 2, SART TN 006/2005 (2005)
11. Vick, A.R., et al.: Comparisons of Experimental Free-Jet Boundaries with Theoretical Results Obtained with the Method of Characteristics. NASA Langley Research Center, NASA TN D-2327 (1964)

Experimental Study of the Base Flow

A. Henckels and A. Gülhan

Deutsches Zentrum für Luft- und Raumfahrt (DLR)
Abteilung Windkanäle
des Instituts für Aerodynamik und Strömungstechnik
Linder Höhe, D-51147 Köln, Germany
andreas.henckels@dlr.de

Summary

At high altitude, conventional rocket nozzles operate at a non-adapted state. A substantial aerodynamic problem is the interaction between the highly underexpanded plume and the ambient flow field. In order to improve the understanding of this interaction an experimental study on a model representing the base region of a booster configuration has been carried out at the hypersonic wind tunnel H2K. At realistic external flow conditions pressure measurements and high speed Schlieren visualizations were performed to study the flow topology for cold and warm nozzle flow at several pressure ratios. Among others, the study focuses on the identification of shock oscillations. Recorded spectra provide the dominating oscillation frequencies linked to flow conditions.

1 Introduction

The integration of the propulsion component plays a key role in the design process of future launchers. A substantial aerodynamic problem is the interaction between the base flow of the vehicle and the hot nozzle jet. Conventional bell-shaped rocket nozzles of fixed geometry are commonly used to expand the exhaust gas. Due to the permanently changing ambient pressure along the trajectory, these nozzles operate at a non-adapted state most of the time and thus suffer from undesirable dynamic loads and thrust losses.

At high altitudes, the large exhaust plume may induce boundary layer separation at the spacecraft surfaces upstream of the nozzle, leading to a significant rise of the boat tail drag and thermal loads on the base structure. In the low pressure region downstream of the base "reverse jets" of the hot exhaust gas may be generated, which could lead to overheating of the external nozzle surfaces. An "upstream induced boundary layer separation" may also influence the stability and control effectiveness. To the same extend as the base flow and the shock structures are unstable, oscillations, which may stimulate "buffeting", are generated. All these aerodynamic effects have a strong impact on the vehicle characteristics and thereby may affect the feasibility of future launcher projects.

The issue of plume interactions in the base region of space vehicles was brought up after 1950 by NASA. Since that time, this topic was investigated theoretically and experimentally by several authors. Early activities by Love et al. [9] provided

A. Gülhan (Ed.): RESPACE – Key Tech. for Reusable Space Sys., NNFM 98, pp. 20–39, 2008.
springerlink.com

comprehensive wind tunnel data of jet boundaries of expanding free jets. At the end of the 60's, Brewer and Carven [1] performed experiments inside a test cell pointing out a reverse jet forming in the base region of a four-engine clustered nozzle configuration. Experimental studies of the rear flow field of ARIANE 5 were performed in France by Reijasse and Délery [11] in 1994.

ESA concentrated activities on this subject in the frame of the FESTIP program by initiating investigations on an axisymetric model configuration with an exhaust nozzle. Associated to this effort, Rubio, Matesanz et al. [12] emphasized the importance of quasi-analytical and engineering methodologies for the prediction of base flow/plume interactions and successfully compared results of their calculations with CFD calculations and experimental data from ONERA. At the University of Delft, Scarano et al. [13] carried out detailed wind tunnel measurements on the FESTIP model by means of Particle Image Velocimetry at a free stream Mach number of two and a nozzle exit Mach number of four.

Parallel to the advances in CFD, a growing number of numerical and combined analytical/numerical approaches to that subject can be noted [e.g. 6,8,10]. Nevertheless, the complex 3D base/nozzle flow field with its separated flow regions and its several viscous interactions remains a challenging task for CFD simulation. In particular, experiments remain essential to understand physical effects and their impact on the overall design of future launchers. Therefore, the H2K facility of DLR's Wind Tunnel Department Cologne has been upgraded in order to carry out complex base flow simulations. The specific test facility capabilities and a long term experience in technology orientated hypersonic research on aerodynamic propulsion components, like inlets and SERN-nozzles [e.g. 2, 3] are the basis of a qualified DLR contribution in that field.

In order to improve the understanding of the interaction of the nozzle flow with the base flow an experimental study on a generic model has been carried out at hypersonic wind tunnel H2K. For experimental simulations a scaled model representing the base region of the Liquid Fly-Back Boosters (LFBB) configuration [15, 16], which is one of the possible future booster options for Ariane, has been designed. To distinguish between different physical effects, the geometry of the model was kept as simple as possible, i.e. as a single nozzle configuration.

This paper covers a description of the experimental set up including the complex internal model design. After a preliminary test campaign [4], runs with a cold and warm nozzle flow were performed at several pressure ratios and at realistic external flow conditions. Wall pressure measurements and Schlieren visualizations were combined to study the flow topology. Among others, this study focuses on identification of pressure oscillations as evaluated from high speed Schlieren visualizations. Further tests with different exhaust gases as Helium and Argon provide information on the influence of the specific heat coefficient and the temperature itself.

2 Experimental Techniques

2.1 Test Facility and Similarity Considerations

In order to simulate the flow field around a scaled afterbody/nozzle model, the conventional hypersonic blow down wind tunnel H2K at DLR Cologne is used. This facility features a free stream test section and an arrangement of eight electrical heaters of 5 MW total power in order to study high temperature effects and to avoid condensation of the test gas around the model. The complete heating system has its own control unit, linked to the main control unit of H2K.

The test flow is generated by contoured Laval nozzles for Mach numbers of 4.8, 5.3, 6.0, 7.0, 8.7 and 11.2. Different Reynolds numbers can be adjusted by the variation of the stagnation conditions. In order to generate an established flow field, the pressure in the test chamber is decreased by a vacuum sphere. During the test run of several seconds, the pressure inside this sphere rises, until the flow breaks down at a certain pressure ratio.

To heat the secondary flow of the base model nozzle, the H2K was upgraded by an auxiliary electrical resistance heater, installed on the floor parallel to the diffuser of the facility (Fig. 1). The electrical power of 260 kW is sufficient to heat air at mass flow rates of 0.5 $kg \cdot s^{-1}$ up to 700 K at pressures of about 20 bars.

To guarantee a reliable operation and to control the performance of the heater, sensors are integrated which measure internal temperatures at sensitive locations. Air mass flow, static pressure and static temperature at the heater exit are measured and transmitted to the main operation desk (Fig. 2). Finally, the heated air is injected directly into the wind tunnel model. The whole piping from the heater through the test chamber wall is insulated to reduce heat losses.

In order to simulate the under-expanded operation of a VULCAIN 3 nozzle, the test condition should match the high altitude condition at a typical point of the ascent trajectory of the LFBB reference concept as close as possible. As reference, the condition 138 s after lift off and shortly before booster separation was chosen. The

Fig. 1. Heater for the nozzle air flow and isolated pipe connected to the H2K test section

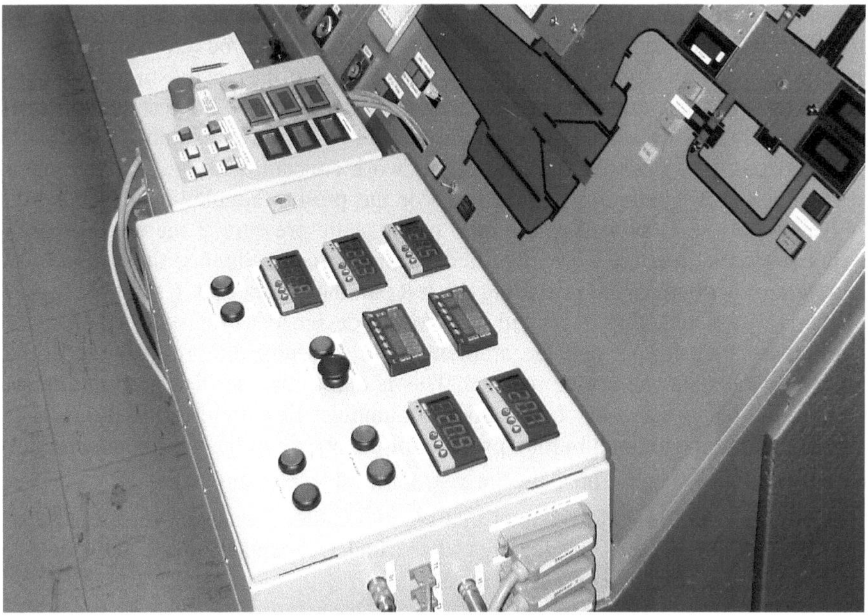

Fig. 2. Control panel for the nozzle flow integrated into the operation desk of the facility

Table 1. Reference condition of the free stream flow

Altitude [km]	50
Mach number	5.3
Flight velocity [$m \cdot s^{-1}$]	1750
Static pressure [bar]	$0.76 \cdot 10^{-3}$
Static temperature [K]	271
Reynolds number [m^{-1}]	$1 \cdot 10^{5}$

characteristic parameters of that condition, i.e. at a flight altitude of 50 km and a flight Mach number of about 5.3, are given in Table 1.

From an aerodynamic point of view, a precise experimental simulation with scaled models requires to reproduce the most relevant similarity parameters. For instance, Mach and Reynolds number govern the viscous flow effects like boundary layer transition or shear layer establishment.

The flight Mach number can simply be reproduced by use of the adequate wind tunnel Laval nozzle. Other important similarity parameters determining the downstream expansion of the plume are the ratio of the nozzle exit pressure to the ambient static pressure as well as the angle of contour divergence at the nozzle exit. Also the momentum ratio between the ambient flow and the gas flow at the nozzle exit should be identical, as this ratio affects the momentum exchange between both sides of the plume shear layer.

Obviously, it is not possible to simulate all these parameters in ground testing, unless identical free stream and exhaust gas temperatures can be realized. In this case the stagnation temperature of more than 3000 K inside the combustion chamber of VULCAIN 3 has to be maintained for several seconds inside the test section of the hypersonic blow down tunnel. This would require major modifications to the facility and its operating procedures. Therefore, this work concentrated on only reproducing the most relevant simulation parameters. For the present simulation, the exit Mach number of the nozzle as well as the ratio of the static pressure at the nozzle exit and the ambient static pressure were identified. Both parameters implied the design of two different nozzle contours [5] providing the test parameters according to Table 2.

It has to be noted, that even with the nozzle designed to reproduce the pressure ratio between nozzle exit and the ambient static pressure, the pressure ratio of the VULCAIN 3 nozzle can not be reached. This is due to the fact that a minimum total pressure of 3 bar is necessary to establish the ambient flow field inside the H2K test section, and that the maximum total pressure of the nozzle of the wind tunnel model is limited to about 20 bar. Thus, the largest achievable pressure ratio p_{nozzle} / p_{∞} of about 100 by nozzle 1 is lower than that of VULCAIN 3 nozzle. Nevertheless it is expected, that significant flow features of the under-expanded nozzle flow field are already evident at this lower pressure ratio.

Table 2. Reference condition of the nozzle flow

Exit section	VULCAIN 3	Nozzle 1	Nozzle 2
Mach number	3.9	3.13	3.88
Expansion ratio ε	35	5	10
Nozzle exit diameter [mm]	1619.8	33.5	47.3
Angle of flow divergence [°]	10.1	10.2	10.0
Total pressure [bar]	139	20 (max.)	20 (max.)
Pressure ratio p_{nozzle}/p_{∞}	401	101	43

In addition to Mach number and pressure ratio, the total temperature of the nozzle flow is another important simulation parameter. This temperature affects the density of the nozzle jet as well as the viscosity at the shear layers. To understand this influence on the base flow field better, the test matrix covers runs with heated exhaust gas. Further runs also cover the use of different exhaust gases as Argon and Helium in order to distinguish between the influence of the specific heat coefficient and the temperature itself.

2.2 Wind Tunnel Model and Measurement Technique

The model design in CATIA [7] was based on two guide lines: On one hand, the external model geometry should resemble typical launcher geometries like the LFFB configuration. On the other hand, the external geometry should be as simple as possible in order to support the CFD mesh generation process and to distinguish between physical effects more easily. In addition, the internal model design should compromise several test requirements, e.g. a modular design and a maximum amount of instrumentation.

As these tests aim at a simulation of the flow downstream an underexpanded nozzle at high altitude, one nozzle with an adequate mass flow is sufficient to generate relevant plume phenomena of the nozzle cluster [14]. Figure 3 shows the resulting shape of the model with this single nozzle, which is the basis for all accompanying CFD activities.

The front section of the model consists of a 36°-cone. In order to demonstrate the influence of the entropy layer development on the establishing boundary layer, tips of different nose radii can be fixed to this cone. A Pitot probe integrated into the nose section allows to measure the free stream condition of the ambient flow. Four circumferential probes at the cones surface support the exact alignment of the model relative to the main flow direction.

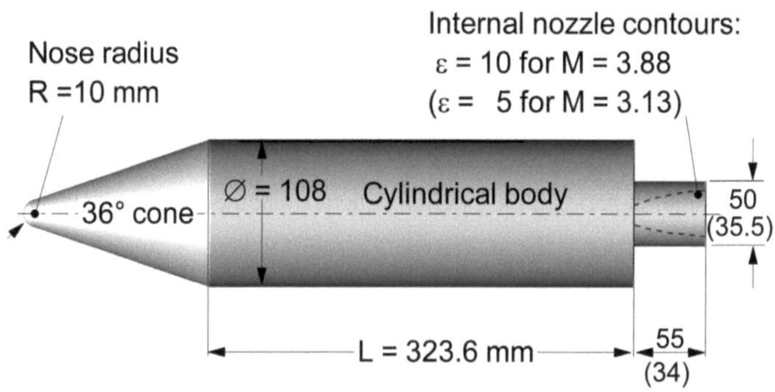

Fig. 3. External shape and dimensions of the wind tunnel model (dimensions for nozzle ε = 5 in brackets)

The length of the adjacent cylindrical model section is 319 mm, so that the boundary layer develops over a total length of about 0.5 m, before it separates at the base shoulder. For the nominal test flow condition (Mach 5.3, $p_0 = 3$ bar and $T_0 = 600$ K) the Reynolds number of about $1 \cdot 10^6$ guarantees laminar separation at that location. This supports the comparison with prospected results from numerical simulations.

The most challenging design element of the model is the air supply, which is needed to feed a sufficient mass flow of hot air into the model. Inside the test section, the cylindrical part of the model is attached to a profiled sting, which contains two parallel ducts to minimize external flow interferences.

Starting from cold test condition, the nozzle flow has to be heated up to the desired temperature level. In order to keep the pressure level in the vacuum sphere, i.e. in the test chamber low enough, which is necessary for a reasonable testing time, the equilibrium nozzle flow conditions have to be established within a short time period.

To meet these requirements, the thermal losses as well as the heated structural mass have to be as small as possible. Therefore, the design philosophy was to separate the cold external model structure, containing the pressure instrumentation, from the hot pressurized internal components, like ducts, settling chamber and nozzle. As shown in Fig. 4, the stilling chamber and the nozzle are supported by ceramic rings inside the cold structure, which are attached to a spring to compensate the thermal extension.

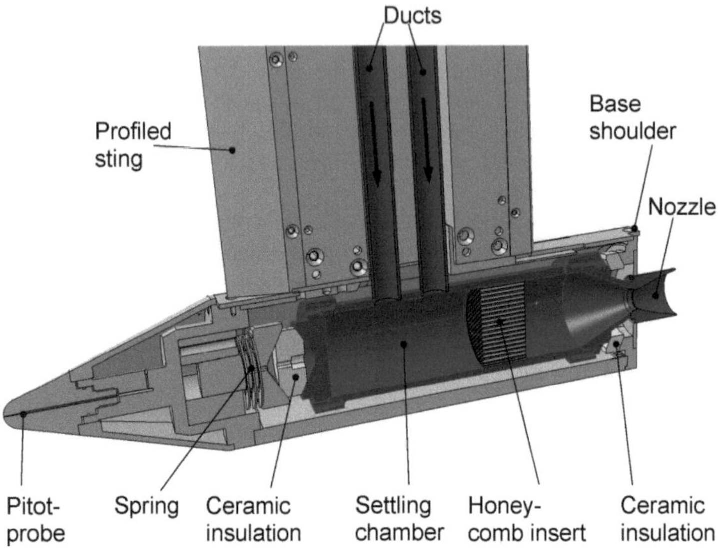

Fig. 4. Thermal insulation between the external structure and the hot internal components of the model

The stilling chamber and an integrated honeycomb insert are used to reduce the turbulence generated by the manifold of the air supply. The maximum pressure inside this chamber is limited to 20 bar. Downstream of the insert, the stagnation condition of the nozzle flow is measured by a thermocouple and a Pitot probe.

Fig. 5. Model installed inside H2K test section

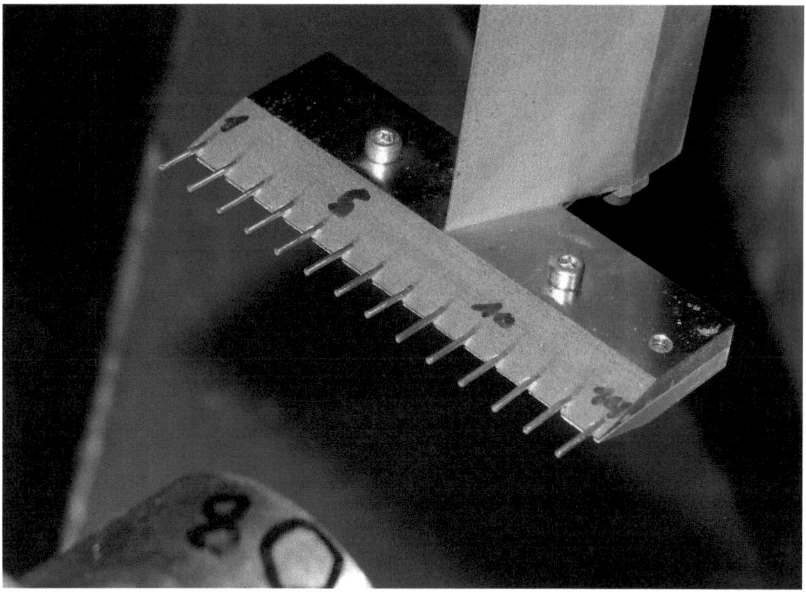

Fig. 6. Rake with 14 Pitot probes of 1 mm diameter downstream the nozzle

Contrary to the bell shaped external contour of real rocket nozzles, the model nozzles are designed with a cylindrical external contour in order to allow the installation of pressure sensors in the nozzle wall. Further pressure sensors are integrated in the model base and in the base shoulder to gather detailed information on the external base flow.

In Fig. 5 the model installed in the test section of the H2K facility is shown. During a run, flow establishment around the model can be monitored by sensitive coincidence Schlieren optics. Parallel, image sections are recorded by a high speed camera (PHOTRON, Ultima APX-RS) at frequencies up to 20 kHz. At selected locations contrast fluctuations taken from these scenes are evaluated by FFT analysis in order to detect flow oscillations.

Information about the entire stationary Pitot pressure distribution downstream the nozzle is gained by a Pitot rake equipped with 14 pipes of 1 mm diameter (Fig. 6). To support assumptions about the wall temperature condition for numerical simulations, the temperature on the model surface is recorded by an infrared camera system (AGEMA, ThermaCAM SC3000 NTS) at 60 Hz.

3 Results

3.1 Wall Pressures and Temperatures

In order to break down the flow field features to a physical interpretation, defined test cases with a comprehensive set of flow data are required. For a definition of the boundary conditions of later CFD simulations the model wall temperature, the

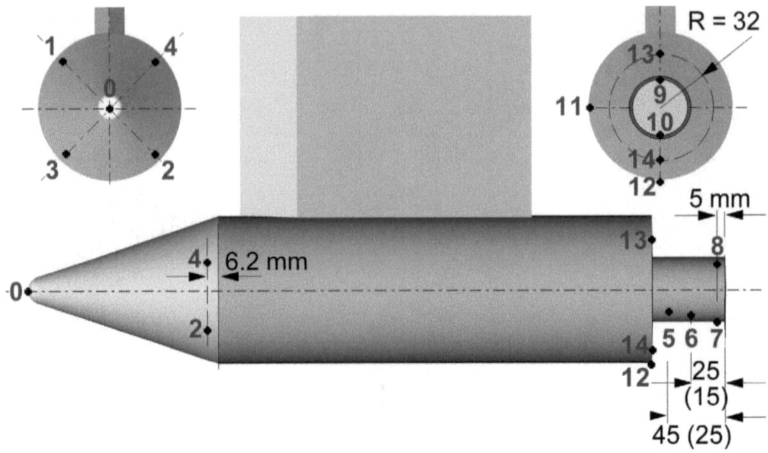

Fig. 7. Location of the pressure orifices for nozzle ε = 10 (nozzle ε = 5)

condition of the incoming flow as well as pressure coefficients at certain model positions (Fig. 7) are required.

For selected test cases the pressure coefficients are listed in Table 3. As can be seen by the small deviation of about 3% of the pressure coefficients at positions no. 1 to 4, the alignment of the model to the oncoming wind tunnel flow was nearly perfect. The interpretation of the pressure coefficients no. 5 to 8, measured on the external nozzle contours, naturally suffers from the relatively low pressure levels in this region. At this point the interpretation of the base flow topology will be supported by numerical analysis later.

Evaluating the differences between pressure coefficients no. 7 and no. 8 measured at opposite positions, an influence of the sting might be concluded. This sting influence is also confirmed by pressure coefficient no. 14, which is lower in comparison to pressure coefficient no. 13. Note that the measured pressure levels are relatively low and thus are subject to a high sensitivity to the model geometry.

For a final judgment of the sting influence, Pitot pressure surveys of the flow field downstream of the model will be shown later. Nevertheless, it is expected that the measured pressure coefficients at the lower model side are not affected by the sting influence.

The flow inside the stilling chamber of the model is reduced to relative low flow velocity (Mach 0.03). Therefore, and because of the installed honeycomb insert, a rather homogenous flow field entering the nozzle is expected. In contrast with this, a significant difference in the pressure coefficients no. 9 and no. 10, measured at opposite sides inside the nozzle near its exit plane, became obvious. These differences correspond to Mach number deviation in the order of about ±1%, which is still acceptable.

A detailed check of the internal nozzle contour by precise measurement instrumentation showed, that the nozzle geometries were nearly axis symmetric. Additionally, enlarged Schlieren images proved a nearly perfect symmetrical nozzle flow pattern.

Table 3. Flow conditions and pressures coefficients for selected runs

Run no.	1	2	3	4
	Free stream condition:			
M_∞	5.27	5.28	5.28	5.27
Re_∞ [10^6 m^{-1}]	2.82	3.05	3.18	2.86
$P_{0\infty}$ [bar]	3.24	3.44	3.44	3.25
$T_{0\infty}$ [K]	584	578	561	581
	Nozzle flow:			
ε	10	10	5	5
Test gas	Air	Air	Air	Argon
M_{exit}	3.88	3.88	3.13	3.81
$T_{0nozzle}$ [K]	290	710	292	292
$Re_{Dnozzle}$ [10^6]	4.7	1.3	4.8	6.2
p_{nozzle}/p_∞	34.8	33.4	94.7	30.0
p_{Pitot}/p_0	0.051	0.051	0.051	0.051
	Pressure coefficients:			
cp_1	0.214	0.223	0.218	0.216
cp_2	0.221	0.230	0.213	0.212
cp_3	0.217	0.225	0.209	0.207
cp_4	0.214	0.219	0.217	0.214
cp_5	0.009	0.011	0.006	0.006
cp_6	0.008	0.008	0.007	0.008
cp_7	0.010	0.003	0.004	0.007
cp_8	0.008	0.016	0.010	0.010
cp_9	3.162	2.968	7.650	3.182
cp_{10}	2.885	2.746	7.544	3.163
cp_{11}	0.005	0.023	0.000	0.003
cp_{12}	0.001	0.010	0.006	0.006
cp_{13}	0.004	0.018	0.010	0.008
cp_{14}	0.005	0.016	0.008	0.005

CFD simulations require assumptions about the temperature condition at the model surfaces, i.e. an adiabatic or isothermal wall at a certain temperature level. The wind tunnel model is machined from steel, having a significant high thermal conductivity. During the entire test run, a nearly homogenous surface temperature on particular model parts is found from infrared images.

The wall temperature of the cylindrical model is mainly influenced by the stagnation temperature of the ambient flow. The wall temperature of the nozzle results mainly from internal heat conduction, and therefore is controlled by the stagnation temperature of the nozzle flow. An evaluation of exact temperatures requires knowledge about the surface emissivity, which for the oxidized steel surface is specified near to 0.65. In Table 4 measured temperatures are referred to relevant stagnation temperatures.

Table 4. Infrared measured model wall temperatures for run no. 2 of Table 3 (emissivity 0.65)

Model component	Test duration [sec]	T_{wall} [K]	$T_{0ref.}$ [K]	$T_{wall}/T_{0ref.}$
Cylindrical body	1	318	$T_{0\infty} = 578$	0.55
	20	323	$T_{0\infty} = 578$	0.56
Nozzle	1	436	$T_{0nozzle} = 710$	0.61
	20	455	$T_{0nozzle} = 710$	0.64

3.2 Base Flow Topology and Sting Influence

As the vehicle traverses through the atmosphere, the ambient pressure decreases to low levels. At static pressures corresponding to 50 km altitude, the model nozzle operates at strongly underexpanded condition, which leads to the flow topology sketched in Fig. 8:

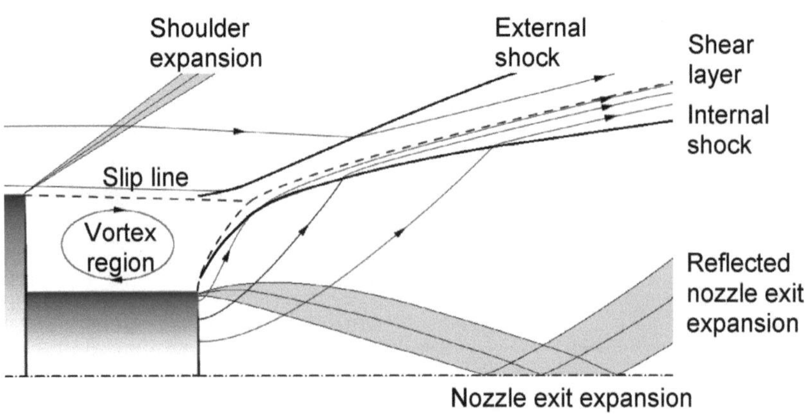

Fig. 8. Sketch of the flow structure downstream the model

At the rear of the cylindrical model, the boundary layer separates as free slip line, forming a small expansion at the base shoulder. This slip line encloses the low pressure base flow region with embedded subsonic vortices.

Further downstream, near to the plane of the nozzle exit, the slip line interacts with the viscous shear layer, which separates the exhaust gas from the ambient flow. Thereby, an external shock is generated, spreading circularly around vehicles rear.

At the nozzle exit, the exhausted high pressure gas rapidly expands by generating an expansion fan spreading towards the plume axis. Thereby, particularly at the edge of the nozzle exit, the flow is turned by large angles (see stream lines). In order to adapt the direction of the core flow to the direction of the viscous shear layer, which is positioned by the pressure balance, an "internal shock" still embedded inside the plume gas is generated. Depending on a sufficient density gradient, which increases by the stagnation pressure of the nozzle flow, the Schlieren images Fig. 9 visualize significant plume features.

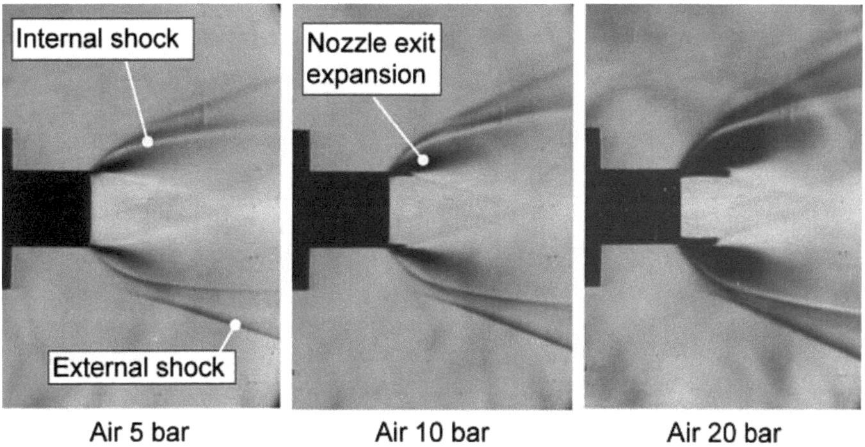

Fig. 9. Schlieren visualization of the air plume at different levels of stagnation pressure. (Nominal external flow condition: Mach 5.3, $p_0 = 3$ bar and $T_0 = 600$ K)

The influence of the ambient flow field on the plume size is stated by runs with the tunnel flow turned off and on. Figure 10 shows the corresponding Schlieren images taken at different stagnation pressures of the nozzle (expansion ratio 5), but at constant nominal conditions of the wind tunnel flow. Since only the stagnation pressure of the nozzle flow, and thus the mass flow through the model is increased, the size of the nozzle plume grows.

The model is mounted inside the test section on a profiled sting, which may disturb the external flow field downstream. The arrangement in Fig. 11 of the measured Pitot pressure contour and a Schlieren image of the base flow field visualizes the small

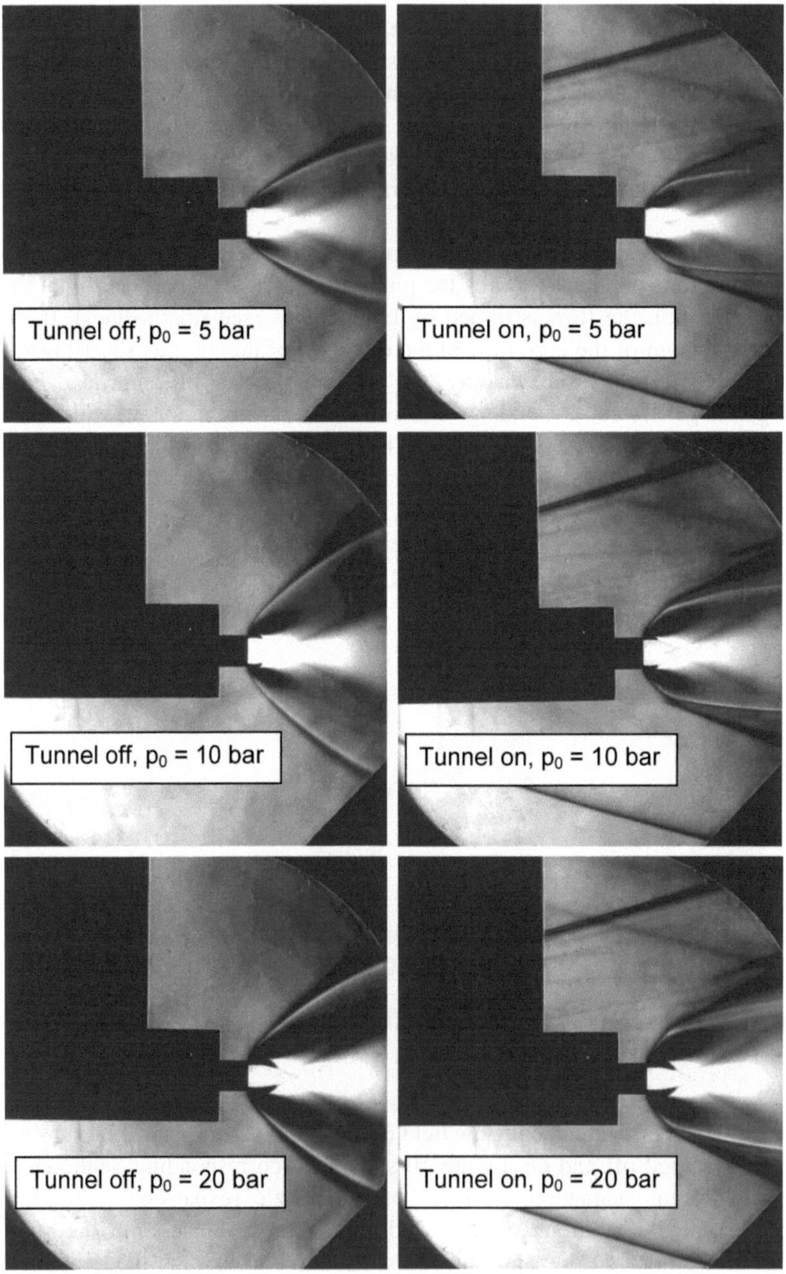

Fig. 10. Schlieren visualization of the air plume at different levels of stagnation pressure (Nominal external flow condition: Mach 5.3, $p_0 = 3$ bar and $T_0 = 600$ K; nozzle $\varepsilon = 5$)

influence of the sting wake on the deformation of the circular external shock, as recognized by the foot print in the pressure contour. Therefore, further relevant flow measurements and high speed flow visualizations focus on the lower section of the model, where no sting disturbances are detected.

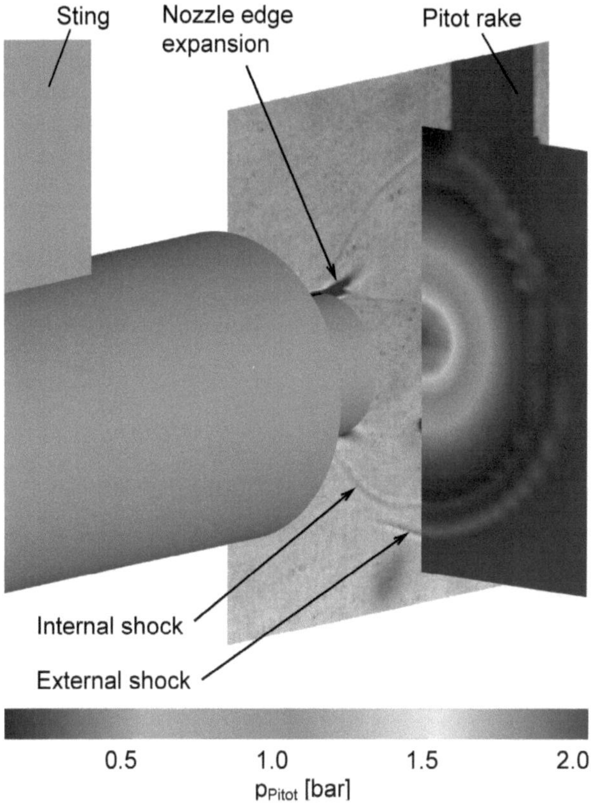

Fig. 11. Arrangement of Pitot pressure contour and Schlieren image of the base flow field

For the LFBB configuration at 50 km altitude the Reynolds number at the shoulder of the cylindrical base section is about $4 \cdot 10^6$, which leads to a less defined transitional character of the boundary layer. Compared with this, the nominal free stream test condition (Mach 5.3, 600 K and 3 bar) guarantees laminar separation at this location. In order to indicate influences of the boundary layer state, runs at different free stream Reynolds numbers, but at an identical pressure ratio between the static pressure at the nozzle exit and the ambient pressure are performed. Nevertheless, at different Reynolds numbers the separating boundary layer leads to an identical afterbody flow field with the same plume size.

3.3 Effects of Different Exhaust Gases on Flow Interactions

Compared to the exhausted air in the experiment, the propelling gas of real rockets is different particularly with respect to establishing temperatures and hence the gas property. To gain basic information about the influence of the gas property on the afterbody flow interactions and to distinguish between the influence of the specific heat coefficient and the temperature itself, runs with different exhaust gases like Argon and Helium are carried out. Expanding these gases through the nozzle, all flow parameters like velocity, temperature and Mach number are affected by the value of the specific heat coefficient. For example, the nominal nozzle exit Mach number increases from about 3.9 for air to about 5.1 for Helium and Argon. Further relevant flow data are indicated by Table 5.

Table 5. Flow conditions for test with different exhaust gases

	Argon	Helium	Air	Air
Specific heat coefficient	1.67	1.67	1.40	1.40
Gas constant [$J \cdot kg^{-1} \cdot K^{-1}$]	208	2077	287	287
Stagnation temperature [K]	300	300	300	300
Stagnation pressure [bar]	10	10	10	5
Mach number at nozzle exit	5.1	5.1	3.9	3.9
Velocity [$m \cdot s^{-1}$]	528	1668	674	674
Static density [$kg \cdot m^{-3}$]	0.55	0.06	0.35	0.17
Static temperature [K]	32	32	73	73
Static pressure [bar]	0.036	0.036	0.073	0.037

For these runs, the stagnation pressure of the nozzle flow is adjusted to 10 bar in order to resolve small density gradients by Schlieren optics, particularly of the Helium flow. For comparison to tests with air, two strategies may be followed by adapting the stagnation temperature of the nozzle flow: Either the stagnation pressure of air can be identical to the stagnation pressures of the other gases or an identical ratio of static pressure across the plume shear layer can be adjusted. The latter condition leads to equivalent plume sizes.

The design of the nozzle contour for air encounters the elimination of reflected flow characteristics at the internal wall. Characteristic angles resulting from the specific heat coefficient of Helium and Argon do not match to the geometry, originally designed for air. This leads to internal reflections, which may grow to significant shocks, spreading from the nozzle wall forwards the plume axis as visible from Schlieren images.

For an external flow of Mach 5.3, $p_0 = 3$ bar and $T_0 = 330$ K, Fig. 12 shows plumes formed at identical stagnation pressures (air, 10 bar) or at identical static

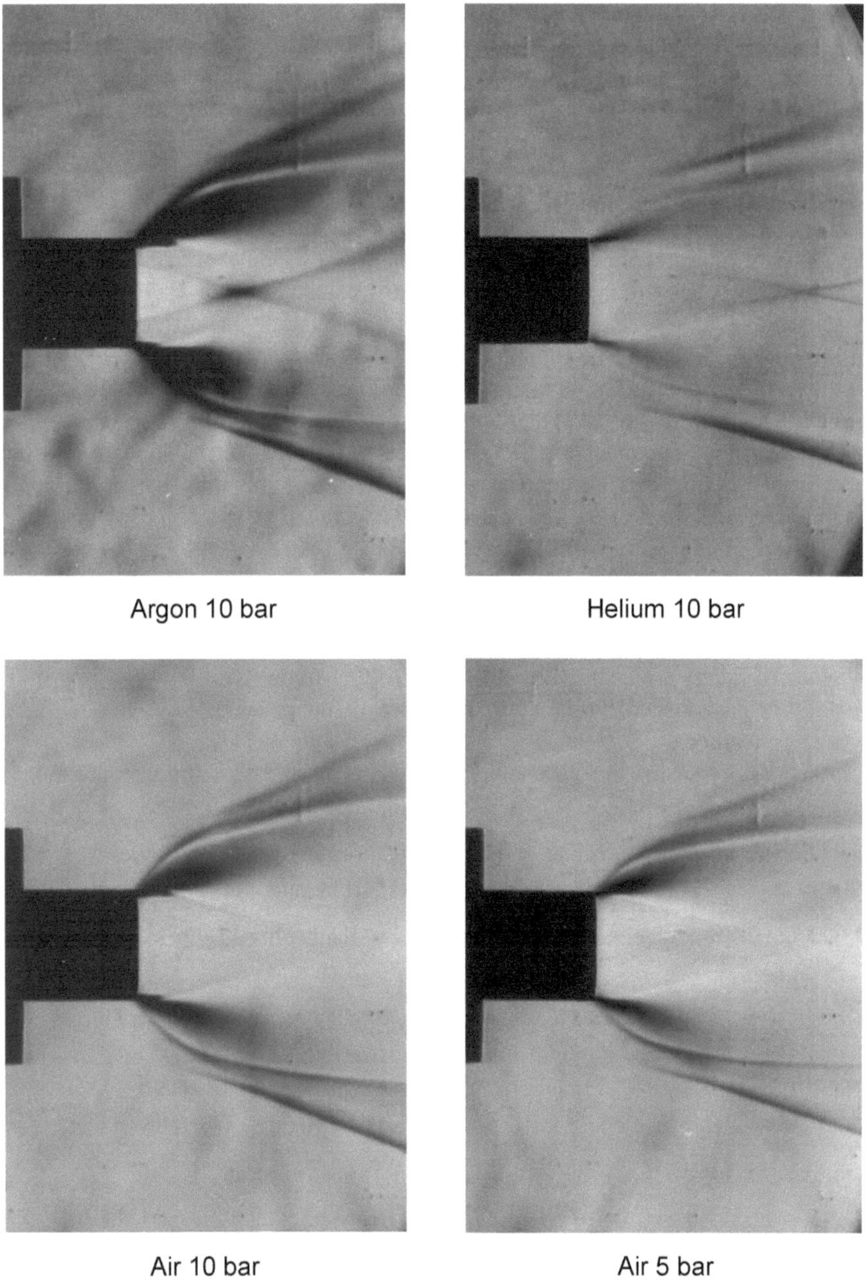

Fig. 12. Argon and Helium plumes in comparison with the air plume according to Table 5

pressure ratios between internal and external flow (air, 5 bar). For the Argon flow, the static temperature reaches a level well beyond the solidification point. Nevertheless, related flow effects could not be observed from the Schlieren images. It is expected, that the solidification process is time shifted due to the low pressure level and the high flow velocity.

3.4 Shock Oscillations and Recorded Spectra

The physical interpretation of time-dependent flow phenomena and its potential on buffeting effects ask for an identification of flow oscillations downstream the nozzle. At selected locations, frequency analyses of the internal and external shocks are performed by evaluating oscillations from high speed Schlieren images (Fig. 13). Thereby, flow conditions are adjusted according run no. 1, Table 4, but for a stagnation temperature of the exhausted air of 660 K.

The measured spectra of Fig. 13 suggest a following interpretation: At position no. 2 and 3 frequencies between around 2 kHz are dominating the oscillation of the internal shock, which is physically linked to the flow condition of the nozzle gas. The amplitude of this oscillation grows for an increasing distance to the nozzle exit.

Marking oscillations of the external shock, which is linked to ambient or base flow field, are found at 5 and 6 kHz at position no. 4 and no. 5. At position no. 1, the influence of the external shock is still weak, so that frequencies between 5 to 6 kHz are slightly pronounced. Further test results with unheated exhaust gas confirmed these interpretations.

In order to transfer the ground testing results to the real flight condition, the reduced frequency $k = 2\pi f D_{ref.} / v_{ref.}$ is computed. A Mach number of 5.3 and a stagnation condition of 3 bar at 600 K leads to a reference velocity, i.e. free stream velocity of $v_{ref.} = 1012$ m/s. Table 6 shows the results for a model diameter of $D_{ref.} = 0.108$ m, associated to the coordinates of the evaluated shocks.

Table 6. Selected locations and results of the frequency analyses

No.	Location	*x* [mm]	*z* [mm]	Frequencies f [kHz]	Reduced frequencies k
1	Internal shock	88	54	5 - 6	3.4 – 4.0
2	Internal shock	108	57	around 2	around 1.3
3	Internal shock	128	61	around 2	around 1.3
4	External shock	93	62	5 - 6	3.4 – 4.0
5	External shock	107	66	5 - 6	3.4 – 4.0

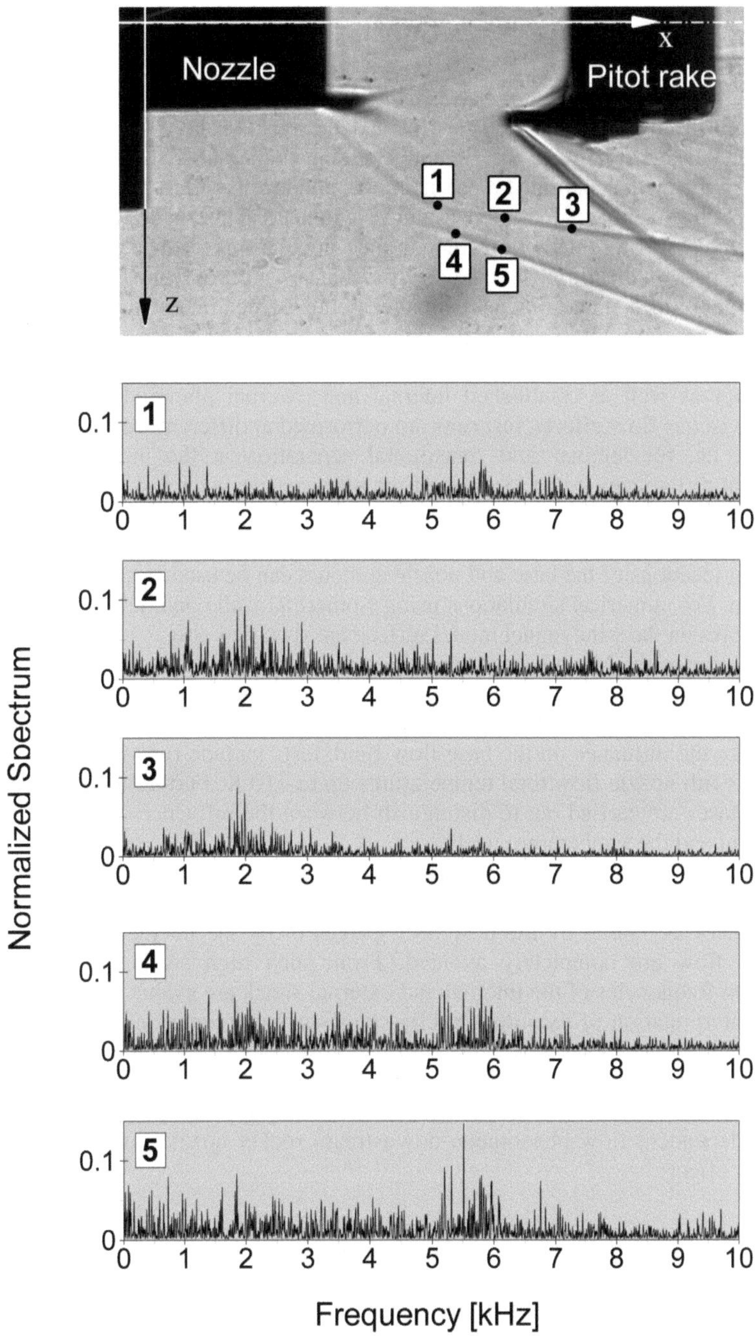

Fig. 13. Frequency analyses of the oscillations of internal and external shock

4 Conclusions

Successful tests on the interaction between the cold air nozzle flow and hypersonic external flow verify the experimental test concept and the functionality of the instrumentation. Negligible deviations between the pressure coefficients measured at different positions on the cone of the model confirmed a perfect model alignment with respect to the free stream direction. Schlieren images indicated an almost symmetrical nozzle flow.

Nevertheless, sting interactions may induce fluctuations of the plume shear layer in the upper part of the flow field. Such disturbances are confirmed by Pitot pressure measurements in a plane downstream the model perpendicular to the mean stream direction. Therefore, all further evaluations focus on the lower part of the flow field.

Schlieren images taken with and without wind tunnel flow visualize the plume shear layer as well as established internal and external plume shocks. In order to indicate viscous flow effects, test runs are performed at different free stream Reynolds numbers, i.e. for laminar and transitional separation at the model shoulder. No significant differences in the location of the plume shear layer and the external nozzle shock are identified.

In combination with these Schlieren images, the measured wall pressure distributions on certain locations of the base and nozzle contours can be used as a reference for CFD validation. For numerical simulations using isothermal wall conditions, the establishing temperatures on the wind tunnel model surface are specified also.

In addition to the Mach number and the nozzle pressure ratio, the total temperature of the nozzle flow is another important simulation parameter. This temperature affects the density of the plume flow as well as the viscosity at the shear layers. To investigate the influence on the base flow field, tests include runs with heated nozzle flow, i.e. with nozzle flow total temperatures up to 710 K. Further tests with different exhaust gases are carried out to distinguish between the influence of the specific heat coefficient and the temperature itself.

High speed Schlieren imaging combined with adequate evaluation software proves to be an efficient non-intrusive measurement technique for frequency analysis. Disturbances as caused by the design of gauges or by the interaction of the gauges with the flow are completely avoided. From such high speed Schlieren images oscillation frequencies of the internal and external shock are evaluated.

The interpretation of recorded spectra suggests that frequencies around 2 kHz are dominating the internal shock, linked to the flow condition of nozzle gas. Marking oscillations of the external shock, which is linked to ambient or base flow, are found between 5 and 6 kHz. These experimental data will support a physical interpretation of time-dependent flow phenomena downstream rocket nozzles and its potential on buffeting effects.

References

1. Brewer, E.B., Craven, C.E.: Experimental Investigation of Base Flow Field at High Altitude for a Four-Engine Clustered Nozzle Configuration. NASA TN D-5164, Washington (1969)
2. Gruhn, P., Henckels, A.: Simulation of a SERN/Afterbody Flow Field regarding heated Nozzle Flow in Hypersonic Facility H2K. In: Breitsamter, C., Laschka, B., et al. (eds.) Notes on Numerical Fluid Mechanics, vol. 78, pp. 236–243. Springer, Heidelberg (2004)

3. Henckels, A., Gruhn, P.: Experimental Studies of Viscous Interaction Effects in Hypersonic Inlets and Nozzle Flow Fields. In: Jacob, D., Sachs, G., Wagner, S. (eds.) Basic Research Technologies for Two-Stage-to-Orbit Vehicles, pp. 383–403. Wiley-VCH, Weinheim (2005)
4. Henckels, A., Gruhn, P., Gülhan, A.: First Experimental Results on the Nozzle/Base Flow Interaction for Launcher Configurations. In: 6th Intern. Symp. on Launcher Technologies, Munich (2005)
5. Herbertz, A., Sippel, M.: Vorbereitung der Untersuchungen zur Heckströmung und Plume-Entwicklung bei Raumtransportsystemen. DLR-IB 647-2004/04, Köln (2005)
6. Houtman, E.M., van der Weide, E., et al.: Computational Analysis of Base Flow/Jet Plume Interaction. In: Proc. of the 3rd European Symp. on Aerothermodynamics for Space Vehicles, ESA SP-426, Noordwijk, pp. 605–612 (1998)
7. Kloft, S.: Auslegung eines Windkanalmodells zur Untersuchung der Interaktion zwischen unterexpandierender Düsenströmung und Raketenumströmung. Diplomarbeit, Fachhochschule Koblenz (2004)
8. Love, E.S., Grigsby, C.E., et al.: Experimental and Theoretical Studies of Axisymmetric Free Jets. NASA-TR-R-6, Washington (1959)
9. Lüdeke, H., Clavo, J.B., et al.: Fluid Structure Interaction at the ARIANE-5 Nozzle Section by Advanced Turbulenc Models. In: Europ. Conf. on Computational Fluid Dynamics, TU Delft, Netherlands (2006)
10. Reig, J.L., Velázquez, A., et al.: Quasi-Analytical Prediction of Base Flow-Plume Interaction. In: Proc. of the 3rd European Symp. on Aerothermodynamics for Space Vehicles. ESA SP-426, Noordwijk, pp. 599–604 (1998)
11. Reijasse, P., Délery, J.: Investigation of the Flow past the ARIANE 5 Launcher Afterbody. J. Spacecraft and Rockets 31, 208–214 (1994)
12. Rubio, G., Matesanz, A., et al.: Prediction of Base Flow/Plume Interaction including 3-D and Real Gas Effects. AIAA Paper 2000-3171 (2000)
13. Scarano, F., van Oudheusden, B.W., et al.: Experimental Investigation of Supersonic Base Plume Interaction by Means of Particle Image Velocimetry. In: Proc. of the 5th European Symp. on Aerothermodynamics for Space Vehicles, ESA SP-563, Noordwijk, pp. 601–607 (2005)
14. Simmons, F.S.: Rocket Exhaust Plume Phenomenology, p. 32. The Aerospace Press, El Segundo, California (2000)
15. Sippel, M., Klevanski, J.: Preliminary Definition of an Aerodynamic Configuration for a Reusable Booster Stage within Tight Geometric Constrains. In: Proceedings of the 5th European Symposium on Aerothermodynamics for Space Vehicles, ESA SP-563, Noordwijk, pp. 21–27 (2005)
16. Sippel, M., Manfletti, C., et al.: Technical Development Perspectives of Reusable Booster Stages. AIAA Paper 2003-7057 (2003)

Large-Eddy Simulation of a Generic Space Vehicle

J.-H. Meiss and W. Schröder

Aerodynamisches Institut
RWTH Aachen
Wüllnerstr. zw. 5 u. 7, 52062 Aachen, Germany
j.meiss@aia.rwth-aachen.de

Summary

The supersonic flow past the blunt base of an axisymmetric aerodynamic body is of fundamental importance for any rocket-like configuration, since these bodies often suffer from high base drag. This work deals with the investigation of the base flow of a cylindrical rocket configuration using large-eddy simulations. In this a system of three vortices in the base region was found to have a major influence on the flow characteristic such as an according decomposition of the pressure on the nozzle wall as well as an upstream directed transport of the hot exhaust gas. A Fourier analysis of the pressure distribution shows a highly time-dependent character of the baseflow and its connection with the interaction of the nozzle flow with the outer flow field. As a prove of the reliability of the used numerical method a comparison with experimental data was carried out showing a good qualitative agreement between the experimental and numerical data.

1 Introduction

1.1 Physical Problem

The supersonic flow past the blunt base of an axisymmetric aerodynamic body is of fundamental importance for any rocket-like configuration. These bodies often suffer from high base drag, which constitutes up to 35% of the overall drag as was shown by Rollstin [18] testing US Army 549 projectiles. In the case of reusable launch vehicles with a larger base area an even greater fraction of the total drag might be reached. The flight regime of current and future launchers always includes subsonic, transonic, supersonic, and hypersonic speeds where the aerodynamic characteristics are strongly influenced by the flow in the base region. For instance, at high supersonic speed in high altitude on the one hand, the base flow is influenced by the outerflow, which separates at the end of the main body, causing the base drag, and on the other hand, the base flow depends on the wide exhaust plume of the underexpanded nozzle flow. This interaction might lead to a boundary layer separation which again rises the base drag additionally. For an overview of the supersonic flow field over a rocket, the Mach number contours are shown. The recirculation area in the bas region is capable to heat up the outer nozzle structure due to a convection of the hot exhaust gas. Similar to the unsteadiness of the flow phenomena like shocks, shear layer, and

A. Gülhan (Ed.): RESPACE – Key Tech. for Reusable Space Sys., NNFM 98, pp. 40–56, 2008.
springerlink.com

expansion waves, the flow in the base region suffers from oscillations, which might lead to critical vibrations. Therefore, it is of high interest to provide accurate analyses of the base flow field and the resulting force acting on the base for the design of future space systems.

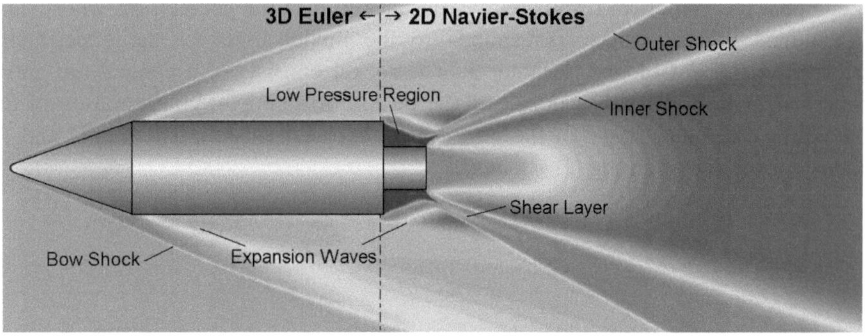

Fig. 1. Mach number contours along a cylindrical Rocket. $Ma_\infty = 5.3$, $Ma_{Noz} = 3.8$. Main body: Results of a 3D Euler computation. Base region: Results of a first 2D Navier-Stokes simulation.

Solely experimental investigations of the flow in the base region are not sufficient to capture the whole flow characteristics, since not all important flow parameters can be satisfied during wind tunnel tests or due to the influence of the support sting, the effect of which on the base flow is not negligible. This leads, for example, to a non-adequate approach to determine the forces acting on the base. Thus, it is useful to combine the experimental approach with a numerical investigation to improve the reliability of both the experimental and the numerical work. In this project the numerical investigation of the base flow is performed using a large-eddy simulation (LES), while the experimental data is provided by DLR Cologne, Chapter 3.

Numerous computational studies have already been done to predict turbulent supersonic base flow of a cylindrical body. Most computational analyses try to predict the supersonic, axisymmetric base flow at $Ma_\infty = 2.46$ as was experimentally investigated in detail by Dutton and Mathur [4] and Dutton and Baswell [5]. The turbulence models range from different Reynolds-averaged Navier-Stokes (RANS) models [3, 15, 20] via DES [10, 12] and LES [7] to DNS [21, 22]. However, the base flow and the base pressure have not always been predicted satisfactorily, especially the RANS models often fail to provide accurate results. Moreover, the DNS is restricted to small Reynolds numbers and to a small integration domain due an extremely high grid resolution. Finally, most of the investigations have mainly been restricted to mean data, while the dynamic problem has barely been tackled. Hence, an LES with a high resolution is used to examine the spatial as well as the temporal behavior of the supersonic base flow.

The article is organized as follows. After a brief description of the problem, the governing equations and the numerical method are discussed in section 3.2 with special emphasis on the formulation of the inflow boundary distributions. Subsequently, in section 3.3 the results of the numerical investigation are presented, including some comparison with experiments, which were carried out by DLR Cologne. Finally some conclusions will close the article.

1.2 Model Geometry and Flow Conditions

To compare the numerical results with experimental data the geometries, whose base is examined in this work, corresponds to the model used for the experiments performed at DLR Cologne, Chapter 3. The rocket consists of a conical top and a cylindrical main body with a nozzle at the base. For a simpler configuration of the base region, the outer shape of the nozzle is cylindrical as well. The model and its dimensions are shown in Fig. 2.

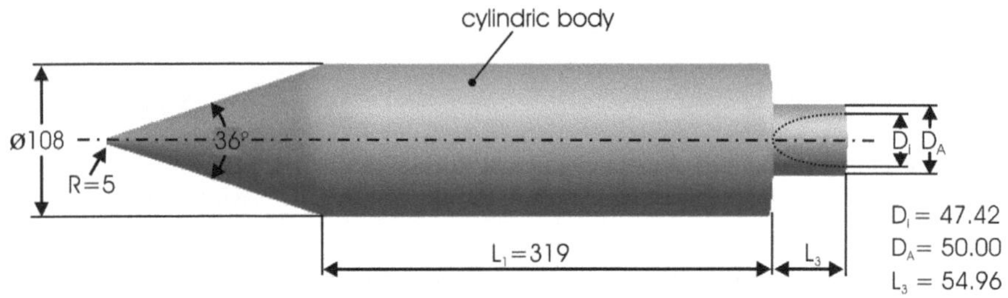

Fig. 2. Geometry of the rocket; measures are in mm

The nozzle expands the exhaust gas to a nominal Mach number of $Ma = 3.8$. Note this value is derived from one-dimensional theory. The real Mach number distribution at the exit is a function of the radius. Since the main task is to investigate the interaction of the nozzle flow and the freestream in the base region, there is no combustion process simulated. Just hot air with a total pressure of 20 bar and a total temperature of 900 K is used as exhaust gas.

The freestream conditions correspond to a Mach number of 5.3 flight at an altitude of about 50 km having a total pressure of 3 bar and a total temperature of 600 K. The main data are summarized in Table 1.

Table 1. Flow conditions at an altitude of 50 km

Flow Conditions		Freestream	Nozzle Flow
Ma	[–]	5.3	3.8
p	[Pa]	$4 \cdot 10^3$	$17 \cdot 10^3$
p_0	[Pa]	$3 \cdot 10^5$	$20 \cdot 10^5$
T	[K]	91	231
T_0	[K]	600	900
c	[m/s]	191	305
c_0	[m/s]	491	601
Re/l	[m^{-1}]	$2.5 \cdot 10^6$	$19.8 \cdot 10^6$

Note, although the freestream Mach number is higher than that at the nozzle exit, the relative velocity of the nozzle flow is positive due to a higher temperature leading to a higher speed of sound.

2 Governing Equations and Solution Technique

To compute the turbulent base flow the three-dimensional unsteady compressible Navier-Stokes equations are approximately solved. In generalized coordinates $\xi_\alpha, \alpha = 1, 2, 3$ these equations read

$$\frac{\partial Q}{\partial 1} + \frac{\partial \left(F_\alpha^I - F_\alpha^V\right)}{\partial \xi_\alpha} = 0 \qquad (1)$$

with

$$Q = \begin{pmatrix} \rho \\ \rho u \\ \rho e \end{pmatrix} \; F_\alpha^I = \begin{pmatrix} \rho u_\alpha \\ \rho u_\alpha u + \dfrac{\partial \xi_\alpha}{\partial \bar{x}} p \\ \rho u_\alpha \left(e + \dfrac{p}{\rho} \right) \end{pmatrix} \; F_\alpha^V = \begin{pmatrix} 0 \\ \dfrac{\partial \xi_\alpha}{\partial x_\beta} \sigma_\beta \\ \dfrac{\partial \xi_\alpha}{\partial x_\beta} \left(u_\alpha \sigma_{\alpha\beta} + \lambda \dfrac{\partial T}{\partial \xi_\alpha} \dfrac{\partial \xi_\alpha}{\partial x_\beta} \right) \end{pmatrix} \qquad (2)$$

where Q represents the vector of conservative variables and F_α^I, F_α^V are inviscid and viscid flux vectors, respectively. The other quantities possess the classical meaning known from the literature.

2.1 Numerical Method

The governing equations for the large-eddy simulation (LES) are the Navier-Stokes equations for compressible fluids, which are filtered according to the size of the grid. Since turbulent flows are characterized by strong interactions between all scales of motion, it is a must to use numerical schemes with only little artificial viscosity. Artificial viscosity considerably impairs the level of energy distribution governed by the small-scale structures and hence falsifies the physical representation of the dynamics of small and large eddies. Thus, in this project a mixed central-upwind AUSM (advective upstream splitting method) scheme with low dissipation and an accuracy of second-order is used. The method has been proven highly accurate in numerous large-eddy simulations [11, 14, 16, 17]. Similar to the monotone integrated LES (MILES) approach [9], there is no explicit subgrid scale (SGS) model implemented, but the inherent dissipation of the numerical scheme serves as a minimum implicit SGS model.

The AUSM method was introduced by Liou and Steffen [13]. Inserting the local speed of sound into the convective expression of the inviscid fluxes after splitting it from the pressure term yields

$$F_\alpha^I = F_\alpha^c + F_\alpha^p = \underbrace{\frac{u_\alpha}{c}\begin{pmatrix} \rho c \\ \rho c u \\ \rho c\left(e + \frac{p}{\rho}\right)\end{pmatrix}}_{f_\alpha^c} + \begin{pmatrix} 0 \\ \frac{\partial \xi_\alpha}{\partial x} p \\ 0 \end{pmatrix}. \tag{3}$$

The numerical flux F_α^c on the cell face, e.g., i ± ½, j, k, reads

$$F_\alpha^c = \frac{1}{2}\left[\frac{Ma_\alpha^+ + Ma_\alpha^-}{2}\left(f_\alpha^{c+} + f_\alpha^{c-}\right) + \frac{\left|Ma_\alpha^+ + Ma_\alpha^-\right|}{2}\left(f_\alpha^{c+} - f_\alpha^{c-}\right)\right]_{i\pm\frac{1}{2},j,k} \tag{4}$$

where the fluxes $f_\alpha^{c\pm}$ and the Mach numbers $Ma_\alpha^\pm$ are determined by left and right interpolated variables obtained using a MUSCLE (monotonic upstream centered schemes for conservation laws) approach for the primitive variables. The remaining pressure term can be calculated by

$$p^\pm = p^\pm\left(\frac{1}{2} \pm \chi Ma_\alpha^\pm\right) \tag{5}$$

with χ representing the rate of change of the pressure ratio with respect to the local Mach number. This quantity determines the numerical dissipation of the scheme. For instance, $\chi = 0$ yields a central splitting. The viscous fluxes are approximated by a general central scheme.

To carry out the integration in time an explicit Runge-Kutta method is applied. To be more precise, a 5-step low storage Runge-Kutta scheme is used the coefficients of which are optimized for maximum stability for a central scheme, α_l = (0.25, 0.1667, 0.375, 0.5, 1). Similar to the spatial approximation the accuracy in time is of the order $O\left(\Delta r^2\right)$ as well. Results of different validation tests are discussed in [14]. In addition, the method was successfully applied to numerous problems including subsonic as well as supersonic flows [1, 2, 6, 11, 16, 17].

2.2 Grid

The distance between the first gridline and the wall is required to satisfy the condition to be less or equal to $y^+ = 1$ to ensure a sufficient resolution of the turbulent near wall flow. In the case of the $Ma = 5$ flow over the rocket the criterion leads to a dimensionless distance of the first gridline of approximately $1 \cdot 10^{-4}$ using the diameter of the main body $D = 108$ mm as reference length. Thus, to minimize the computational effort due to too many grid points, the integration domain is restricted to the base region. Moreover, the core of the grid is designed such that gridlines are diverted to save points within the inner core, where the gradients of the flow parameters are of lower order. The whole grid, which is shown in Fig. 3, consists of $25 \cdot 10^6$ points equally distributed on 48 blocks to allow an efficient concurrent computing via MPI (Message Passing Interface).

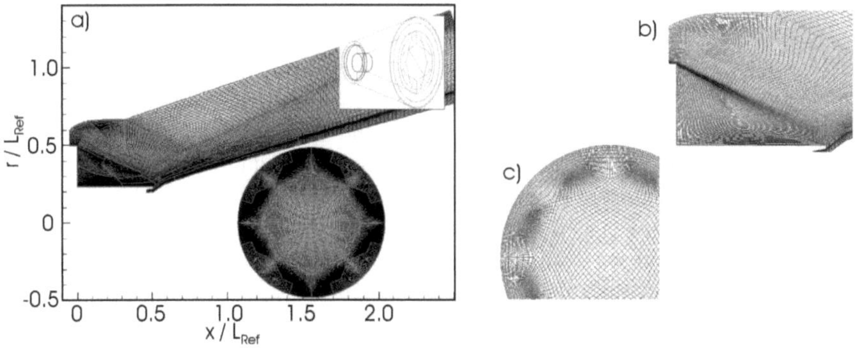

Fig. 3. Grid for the full 3D Navier-Stokes calculations (a). Detailed view of the base region (b) and the core (c). $L_{Ref} = 108$ mm.

To develop the initial distribution for the 3-dimensional flow field, a preliminary 2-dimensional calculation of the base region is performed. After convergence the solution is rotated about the axis of symmetry. Since the structure of the grid defines the quality of the solution, the results of the 2-dimensional flow field are used as a substructure to generate the fully 3-dimensional grid. That is, the grid is generated regarding the direction of the main velocity by considering regions of higher gradients at which the grid has to be clustered, particularly in the boundary and shear layers. Finally, the outer shape of the domain ensures that the shear layers and the recompression shock leave the domain via the downstream boundary.

2.3 Boundary Conditions

Especially the prescription of the flow conditions at the inlet of the base region is of interest. Three formulations are imposed as the inlet boundary to take into account the freestream conditions in the far field, the wall-bounded shear, and the jet of the nozzle. Therefore the set of inflow conditions comprises a fixed prescription for the freestream around the main body, an addition of a turbulent boundary layer, and a predefined outflow of the nozzle. The first condition, the fixed prescription where all conservative variables are provided, is derived from an Euler calculation of the flow over the cylindrical rocket configuration, which generates the high-speed flow field downstream of the main body, influenced by the bow shock and expansion waves.

$$Q = Q_{euler} = const\,. \tag{6}$$

Table 2. Comparison of the static data between the free flow and the flow upstream the base

Variable		Freestream	Baseflow
Ma	[-]	5.3	5.13
ρ	[kg/m³]	0.0155	0.0127
u	[m/s]	1011.56	1006.38
p	[Pa]	402.3	349.3
T	[K]	90.66	95.79
Re/l	$[m^{-1}]$	$2.5 \cdot 10^{-6}$	$19 \cdot 10^{-6}$

The difference of the static values between the freestream and the flow along the main body, which results from the 3-dimensional Euler calculation, is given in Table 2.

To incorporate the viscous effects, the second part of the inflow condition is introduced along the wall by generating a turbulent boundary layer. To estimate its thickness δ some assumptions are made. It is assumed that the development of the boundary layer matches that of a flat plate. It is fully turbulent at the intersection of the cone and the cylinder. In doing so the thickness of the boundary layer at the inlet of the domain can be estimated according to Schlichting [23].

$$\delta(x) = 0.37x\left(\frac{u_\infty x}{\nu}\right). \tag{7}$$

Here x denotes the length of the cylinder (= 319 mm), u_∞ is the velocity determined by the Euler computation, and ν denotes the dynamic viscosity of air ($= 5 \cdot 10^{-4}$ m²/s).

The boundary layer velocity distribution is approximated by the 1/7-law

$$\frac{u}{u_\infty} = \left(\frac{y}{\delta}\right)^{\frac{1}{7}} \tag{8}$$

where y represents the wall-normal distance. The temperature distribution is given by assuming $T = T(u)$, i.e.,

$$\frac{T}{T_\infty} = \left(1 - \frac{T_W}{T_\infty}\right)\frac{u}{u_\infty} + r\left(\frac{\gamma - 1}{2}\right)Ma_\infty^2\left(1 - \frac{u}{u_\infty}\right)\frac{u}{u_\infty} + \frac{T_W}{T_\infty} \tag{9}$$

where T_∞ and Ma_∞ represent the static temperature and the Mach number of the Euler flow, respectively, $T_w = 300$ K denotes the temperature at the wall, $r = 0.88$ indicates the recovery factor, and $y = 1.4$ describes the ratio of specific heats for air. Taking into account the normal momentum boundary layer equation to determine the pressure distribution the density profile can be computed via the perfect gas law. Since only time-averaged quantities are prescribed the inflow condition can be written

$$Q = Q_{bound} = f\left(\frac{y}{\delta}\right). \tag{10}$$

Finally, the third inflow condition describes the outflow of the nozzle. Here the time average data is taken from a solution of the nozzle flow provided by the Techn. Univ. of Munich, see Chapter 5.

$$Q = Q_{nozzle} \, . \tag{11}$$

Since the nozzle flow is computed using the Navier-Stokes equations the data already contains a boundary layer.

At the rigid walls of the rocket the no-slip condition and a constant temperature of 300 K are imposed. Since supersonic experiments only allow a limited measurement period, the isothermal wall condition is a good approximation of measurements. Again, the pressure is computed by assuming the wall-normal gradient to vanish

$$v = 0 \quad T_W = constant \quad \frac{\partial p}{\partial n} = 0 \, . \tag{12}$$

Note the nozzle flow presented in chapter 4 is computed using an adiabatic wall condition leading to an inner wall temperature of about 650 K. To ensure a smooth transition the temperature is linearly distributed along the thickness of the nozzle.

Since the flow is supersonic in the streamwise direction at the outflow boundary, no upstream propagating waves occur near the boundary and as such the outflow distribution is completely determined from the interior solution.

3 Results

The results presented in this article are discussed in four sections. Section 3.1 deals with the comparison of two different base flows at different nozzle exit flows, followed by an analysis of the main base flow characteristics in section 3.2. A comparison to experimental data is presented in section 3.3 and next, in section 3.4 the findings of a Fourier analysis of the pressure fluctuation in the base region are described.

3.1 Variation of the Inlet Condition at the Nozzle

In a first step the boundary condition at the inlet was modeled simply by using a uniform Mach number distribution ($Ma = 3.8$) corresponding to the one-dimensional theory with an additional thin boundary layer according to the 1/7-law. When the data of the nozzle flow was available, the rectangular velocity distribution was replaced by the nozzle flow based on the Navier-Stokes solution. Figure 4 gives a comparison of the temperature and the pressure distribution in the base region at two different nozzle flows.

The effect of the parabolic velocity profile is a stronger expansion of the exhaust gas, which widens the subsonic base area. This results in a lower deflection of the main flow and weaker expansion waves at the end of the main body. This again reduces the acceleration of the air resulting in a higher static pressure as well as a higher static temperature compared to the base flow with the rectangular nozzle flow. To be more precise the pressure ratio grows by more than 15% and the temperature

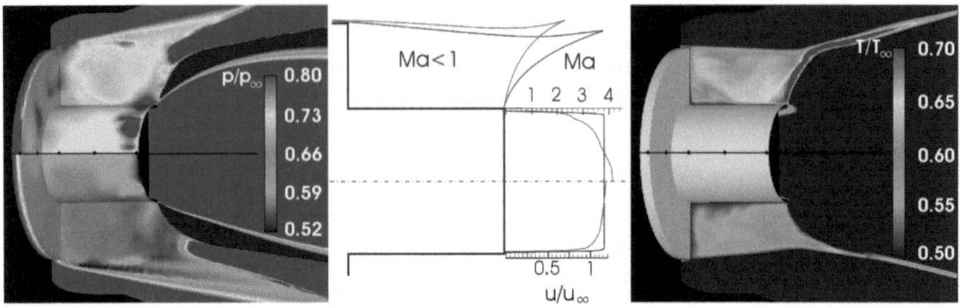

Fig. 4. Pressure (left) and temperature (right) distribution at various nozzle exit distributions; 1D theory (top), Navier-Stokes solution (bottom)

ratio by about 10%. Thus, it is shown that a correct simulation of the nozzle flow has a major influence on the base flow characteristics.

3.2 Base Flow Characteristics

The following results are computed using the Navier-Stokes solution based mean flow distribution at the nozzle exit. The velocity distribution and the streaklines shown in Fig. 5 indicate some turbulent flow structures.

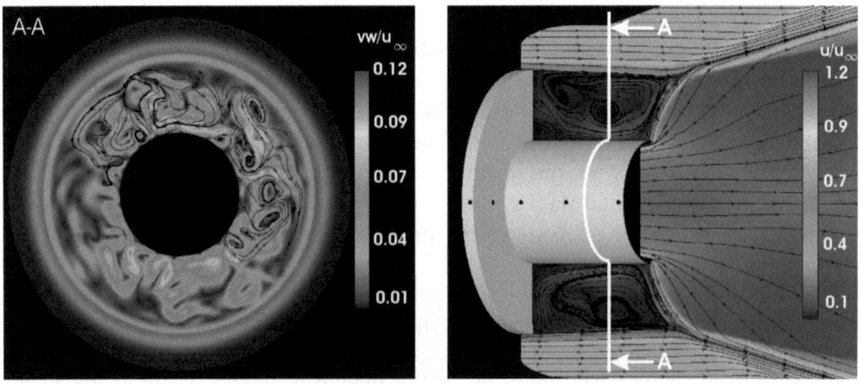

Fig. 5. Instantaneous velocity distribution in the base region, $Ma_\infty = 5.3$, $\alpha = 0°$; radial-azimuthal plane (left), streamwise-radial plane (right)

The fluctuations range from about 1% up to 10% of the freestream velocity in the direction normal to the main flow (Fig. 5, left). There are random structures visible on the whole base, although the rocket base is perfectly symmetric. In the subsonic region in Fig. 5 the velocity possesses values up to 20% of the freestream velocity.

The analysis of the streaklines indicates three zones of the base flow which are characterized by the vortex structure illustrated in Fig. 6. The main vortex covers nearly the whole subsonic region and rotates in the clockwise direction. A small corner vortex is located at the junction of the nozzle and the main body and a free shear layer vortex at the exit of the nozzle between the base area and the nozzle plume. These two side vortices rotate in a counterclockwise direction. The main vortex periodically decompses into two smaller vortices, while the side vortices are stable over the whole period of time. The streaklines shown in Fig. 5 display the main vortex once in its full dimension and once while it is split.

The discussion of the pressure distribution of the base area is divided into the pressure on the outer nozzle wall and the pressure on the base wall of the main body. Since the outer shape of the nozzle possesses a constant radius in the streamwise direction, its pressure distribution has unlike its wall shear stress no direct effect on the base drag. By contrast, the pressure acting on the back main body wall has to be considered completely, while the impact of the wall shear stress distribution on the overall drag can be neglected.

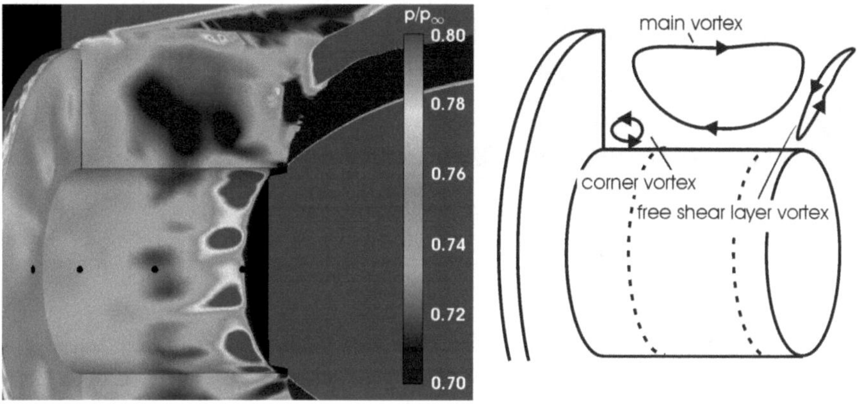

Fig. 6. Pressure distribution on the outer nozzle (left). Schematic of the main vortex distribution (right).

First, let's consider the pressure distribution on the outer nozzle in Fig. 6. Since the pressure is directly connected to the velocity distribution there are similar to the vortex pattern three distinct regions of the pressure distribution. Close to the base wall, the pressure possesses an average value of about 75% of the static freestream pressure. In the second area the pressure drops in certain areas to only 71% of p_∞. At the end of the nozzle, the pressure rises to the highest values in the base area, i.e., approximately 80% of p_∞ are reached. The similarity between the pressure and the velocity distribution is evidenced in the schemtic in Fig. 6. The main vortex is locally of higher speed leading to a lower pressure in the center nozzle area, while the free shear vortex transports fluid onto the wall resulting in a higher pressure.

The pressure distribution on the base wall dominates the base drag. In the following we will analyze its temporal behavior. Figure 7 displays the pressure distribution on the base wall at three different times. The pressure varies between 73% and 78% of p_∞. According to the turbulent velocity structures in the normal direction, the pressure is highly unsteady and nonsymmetrical despite the axisymmetric geometry and flow conditions. While Figs. 7a and 7c show a pressure distribution in the circumferential direction indicating a wave with six and five wave periods, respectively, Fig 7b displays a more arbitrary pressure distribution that cannot be described by a clear wave structure. This nonsymmetric distribution leads to a resulting pressure force vector, which is neither parallel to the axis of symmetry nor does it intercept the axis. Consequently, there will be external forces about all axis, which lead to high frequency oscillations of the rocket.

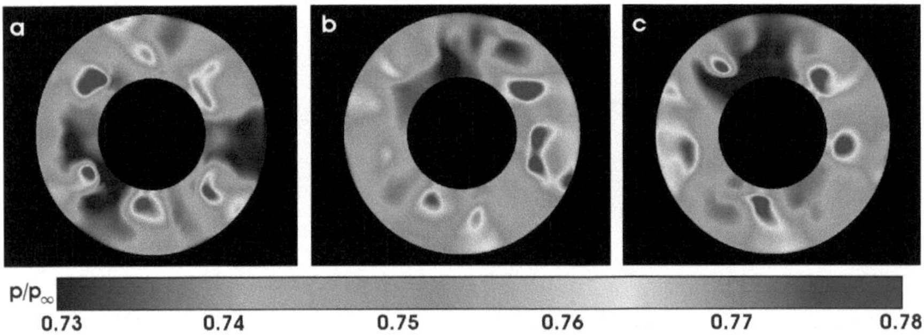

Fig. 7. Pressure distribution on the rocket base, $\Delta t_{ab} = 12 \cdot 10^{-4}$s, $\Delta t_{bc} = 2 \cdot 10^{-4}$s

The vortex structure does not only determine the pressure distribution, but also the temporal distribution of the temperature in the subsonic region. The shear layer vortex distributes the hot exhaust gas in the downstream area of the base region, while the main vortex transports the hot gas into the low pressure area. According to the structure of the main vortex the hot air is directed right onto the rearward nozzle wall, which is heated up significantly before the hot fluid flows in the radial direction along the wall and finally, it is distributed in the whole base region via convective and diffusive effects.

The time dependent process of the transport of thermal energy is shown in Fig. 8, where the temperature distribution is given at equidistant time steps. Note the

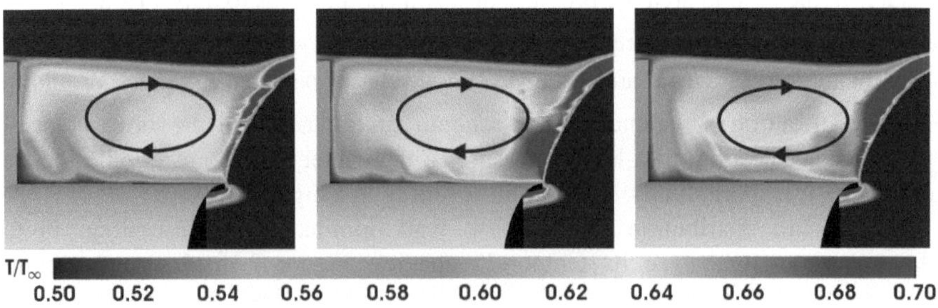

Fig. 8. Lapse of upstream directed heat convection, $\Delta t_{Frame} = 3 \cdot 10^{-4}$s, $T_{Ref} = 600$ K

calculations were performed using an isothermal boundary condition for the wall. Thus, the heating of the wall that will occur under real flight conditions is not simulated. Nevertheless, a reversal transport of the hot exhaust gas with a thermal load on the outer nozzle wall can be predicted.

What are the aerodynamic consequences of the flow characteristics in the base region for the rocket flight? The analysis of the drag coefficient and the momentum coefficient of the base area as a function of time in Fig. 9 indicates the magnitude and the dynamics of the side forces. The drag coefficient contains the integrated pressure force and the integrated shear stress force along the base and the outer nozzle wall. The center of rotation for the moment coefficient coincides with the center point of the main body's base wall.

The drag coefficient in the streamwise direction dominates the overall drag coefficient having a mean value of about -0.0604. The remaining drag coefficients cd_y and cd_z in the directions normal to the axis of symmetry possess a vanishing mean value as it is expected for a symmetric configuration. The axial and the two radial drag coefficients obtain an amplitude of about $1.5 \cdot 10^{-4}$. A Fourier analysis to obtain the frequency spectra is discussed in section 3.4.

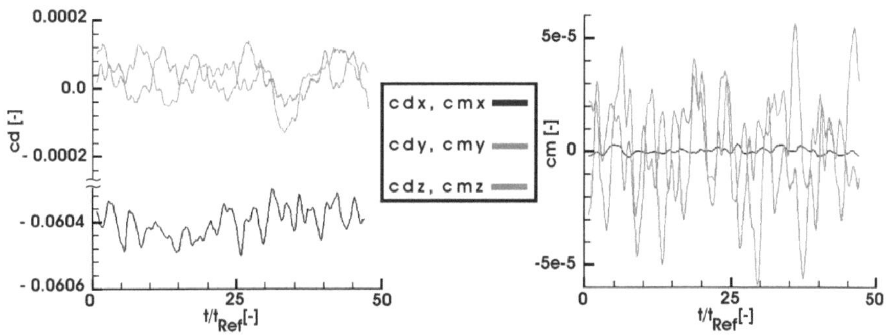

Fig. 9. Time dependent values of drag and moment coefficients, $t_{Ref} = 1 \cdot 10^{-4}$s

The moment coefficients *cm* oscillate about the common mean value $\tau_m = 0$. Since the forces in the radial direction are of much lower intensity, the moment along the main axis can be neglected. The moments along the two radial axes are mainly a consequence of an eccentric center of pressure. Their amplitude is much higher than that of the main axis moment, but having a peak value of about $5 \cdot 10^{-5}$ it is still of low order.

Combining the information about the drag forces and moments the position of the center of pressure on the base wall and the angle of the resulting force vector can be determined. The results are illustrated in Fig. 10. The left diagram shows the time dependent center of pressure, which fluctuates randomly around the center point of the base at a maximum radius of $1 \cdot 10^{-3}$, i. e., 0.1% of the diameter. The angle of the drag force provides a relation between the force in the streamwise direction

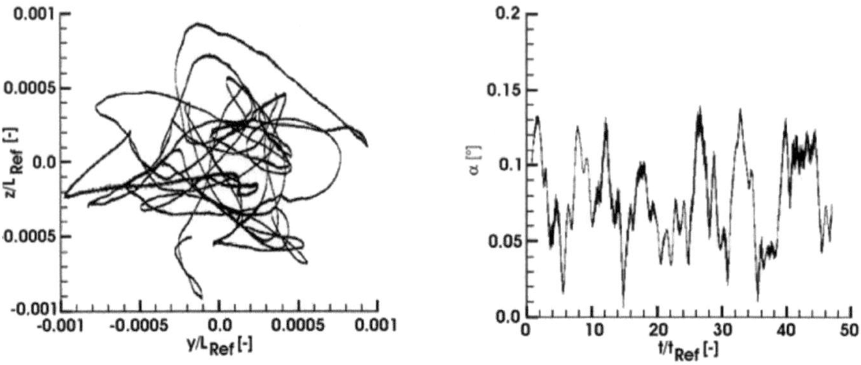

Fig. 10. Fluctuation of the center of pressure at the base for $\Delta t = 5 \cdot 10^{-3}$s, $L_{Ref} = 108$ mm (left). Maximum angle of the force vector, $t_{Ref} = 1 \cdot 10^{-4}$s (right).

represented by cd_x and the radial forces represented by cd_y and cd_z. According to the dominating streamwise force the resulting angle never exceeds 0.2°.

3.3 Comparison with Experimental Data

Next the numerical data will be juxtaposed with experimental findings. To do so, the time dependent pressure coefficient on the wall is compared with the mean experimental data obtained by DLR Cologne, Chap. 3. Pressure measurements were performed at five positions along the axis (Fig. 11). The probes no. 5, 6, and 7 are positioned on the outer nozzle wall, probe no. 12 is located at the end of the main body and probe no. 14 measures the pressure on the base wall.

The flow conditions of the numerical and experimental investigation vary only in the nozzle exit condition. While the pressure ratio at the nozzle exit of the numerical investigation is $p / p_\infty = 42$, the experiments were done at $p / p_\infty = 68$ and $p / p_\infty = 36$. The total temperature at the nozzle exit is 900 K in the numerical simulation whereas it is 300 K in the experiments.

It is expected that according to the varying pressure ratios the computed pressure coefficients will be in-between the experimental mean values. Figure 11 shows that this is true for probes that are located at the base wall, no. 12 and no. 14. The discrepancy as to the pressure coefficients along the nozzle fails, can be explained by taking into account the difference of the total temperature of the exhaust gas. The closer the probe to the end of the nozzle, the mere pronounced the difference between the numerical and experimental finding. As mentioned before, the vortex structure of the base possesses an impact of the hot exhaust gas onto the end of the nozzle followed by an upstream directed transport of the hot gas along the outer wall. This energy transfer causes a larger pressure rise near the nozzle exit and is weakened further upstream, i.e., close to the base.

Overall it can be concluded that, although there is some differences in the experimental and numerical results due to the various nozzle data, the comparison of the data is quite promising.

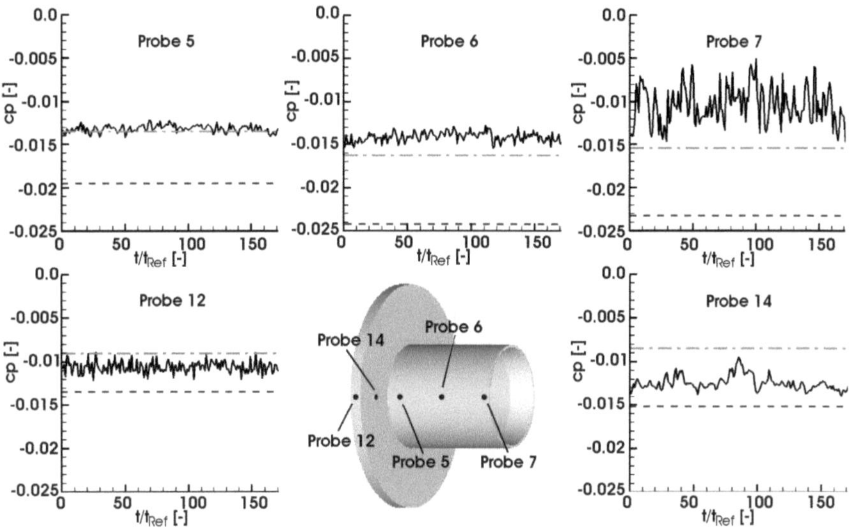

Fig. 11. Comparison of numerical and experimental pressure coefficients
Experiment: — — · Re = 2.6 · 10^6 m^{-1}, p_{noz}/p_∞ = 36, T_{0noz} = 300 K
— · — Re = 2.7 · 10^6 m^{-1}, p_{noz}/p_∞ = 68, T_{0noz} = 300 K
Numerics: —— Re = 2.0 · 10^6 m^{-1}, p_{noz}/p_∞ = 42, T_{0noz} = 900 K, $t_{Ref} = 1 \cdot 10^{-4}$s

3.4 Fourier Analysis

The exact knowledge of the temporal pressure fluctuations are importantfor the design of future launch vehicles. Their periodic character might stimulate the rocket to oscillate possibly leading to critical vibrations and flight instabilities. When the characteristic frequencies and the amplitude of the pressure fluctuations are known measures can be taken to compensate these oscillations.

This section gives a succinct discussion of the Fourier analysis of the pressure distribution at the probes. The results of the analysis are displayed in Fig. 12. It provides the frequency spectra at the probe locations. The affiliated pressure distribution is given in Fig. 11. For the Fourier analysis a discrete Fast Fourier Transformation (FFT) is applied to a set of 2^{19} $(\approx 0.5 \cdot 10^6)$ pressure coefficient values for each probe with an equidistant time step of $\Delta t = 1.3 \cdot 10^{-8}$ s resulting in an overall dimensionless period of time of $\Delta t_{full} / t_{Ref} = 68.2$, where t_{Ref} is $t_{Ref} = 1 \cdot 10^{-4}$ s.

Probes no. 5, 6, and 14, which are located close to the base-nozzle junction, have likewise dominating frequencies. One can be found at 0.3 kHz and the other one at approximately 1 kHz. Moreover, probe no. 14 possesses an additional peak at 0.6 kHz. Note the fluctuation in the base region seems to be linked to the fluctuation of the shear layer. At the experiments the fluctuation of the shear layer had a frequency of 1 kHz.

Probes no. 7 and 12 differ from the base behavior, when a detailed analysis of the data is performed. Probe no. 12, which is at the aft end of the cylinder reveals the

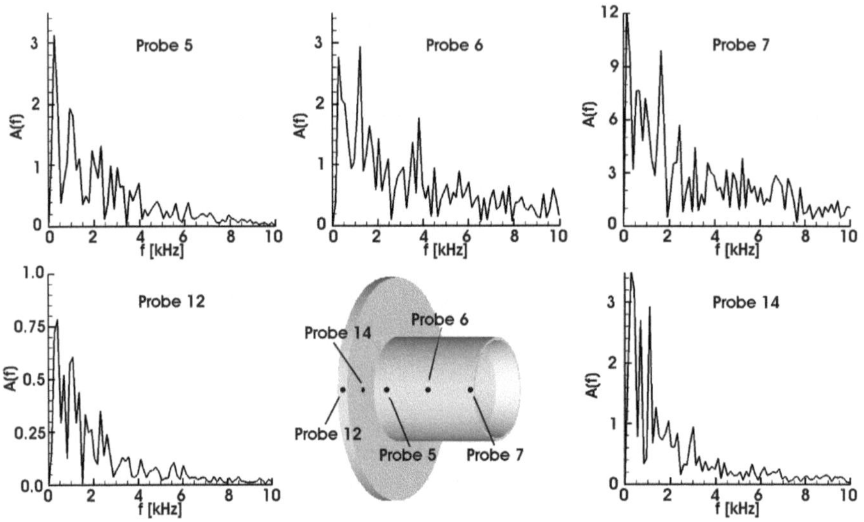

Fig. 12. Fourier spectra of the pressure fluctuations at the locations of the probes

same dominating frequencies as the aforementioned probes, though its amplitude is smaller by roughly a factor of 4. The second probe no. 7 differs due to its location near the exit of the nozzle. The spectrum is influenced by the free shear layer vortex resulting from strong interactions of the nozzle and the base flow. It has already been indicated in Fig. 11 that the amplitude of the pressure fluctuations at this probe is much higher. At this position the main frequencies of the base region, 0.3 kHz, 0.6 kHz, and 1 kHz plus another dominant frequency at 1.6 kHz possess amplitudes which are up to 4 times higher. Besides, there are numerous frequencies up to 8 kHz that have the same amplitude as those in the base region.

4 Concluding Remarks

The base flow of a cylindrical rocket configuration was investigated using large-eddy simulations. A comparison of a rectangular nozzle flow distribution with a parabolic distribution shows quite an influence of the nozzle flow on the base pressure and on the base temperature. This result underlines the correct jet distribution to be very important to accurately predict the base flow.

A system of three vortices was found to have a major influence on the flow characteristic in the base region. They not only decompose the nozzle wall into three pressure regions in the streamwise direction, but they also lead to an upstream transport of the hot exhaust gas near the wall. The resulting heating of the structure could require a cooling system, depending on the total temperature of the exhaust gas.

The investigation of the pressure distribution and the resulting drag force vector shows that the distribution in the circumferential direction is highly time dependent. As expected when a symmetric body is considered, the mean values of the radial forces as well as the moment forces vanish. Furthermore, the maximum eccentricity of the center of pressure and the angle of the drag force vector are only of low magnitude such that their influence on the flight stability is only of low order. Nevertheless, a Fourier analysis of the time-dependent pressure distribution at five positions corresponding to the experimental probe locations were carried out leading to three dominating frequencies in the base area. Experimental data indicate one frequency to be triggered by the fluctuating nozzle. This shows that there is a connection between the subsonic base region and the downstream located interaction of the nozzle flow with the outer flow field, although the main flow as well as the exhaust flow are supersonic.

Finally, the computed data is compared with experimental data. Since the flow conditions of the nozzle differ in pressure and temperature a detailed quantitative comparison cannot be made. However, the juxtaposition show a good qualitative agreement between the experimental and numerical data proving the reliability of the used numerical method.

References

1. Alkishriwi, N., Meinke, M., Schröder, W.: A Large Eddy Simulation Method for Low Mach Number Using Preconditioning and Multigrid. Computers and Fluids 35, 1126–1136 (2006)
2. Alkishriwi, N., et al.: Large-eddy simulations and prticle-image velocimetry measurements of tundish flow. Steel Research International 77, 565–575 (2006)
3. Benay, R., Servel, P.: Two-Equation k-σ Turbulence Model: Application to a Supersonic Base Flow. AIAA Journal 39, 407–416 (2001)
4. Dutton, J.C., Baswell, B.A.: Flow Visualizations and Measurements of a Three-Dimensional Supersonic Separated Flow. AIAA Journal 39, 113–121 (2001)
5. Dutton, J.C., Mathur, T.: Base-Bleed Experiments with a Cylindrical Afterbody in Supersonic Flow. J. of Spacecraft and Rockets 33, 30–37 (1996)
6. El-Askary, W., Meinke, M., Schröder, W.: LES of Compressible Wall-Bounded Flows. AIAA Paper 2003-3554 (2003)
7. Forsythe, J., et al.: Detached-Eddy Simulation With Compressibility Corrections Applied to a Supersonic Axisymmetric Base Flow. J. of Fluids Engineering 124, 911–923 (2002)
8. Fureby, C., Grinstein, F.F.: Monotonically integrated large eddy simulation of free shear flows. AIAA Journal 37, 544–556 (1999)
9. Fureby, C., Grinstein, F.F.: Recent progress on MILES for high Reynolds number flows. J. of Fluids Engineering 124, 848–861 (2002)
10. Fureby, C., Kupiainen, K.: Large-Eddy Simulation Of Supersonic Axisymmetric Baseflow. Turbulent Shear Flow Phenomena, TSFP3, Japan (2003)
11. Guo, X., Meinke, M., Schröder, W.: Large-Eddy Simulation of Film Cooling Flows. Computers and Fluids 35, 587–606 (2006)
12. Kawai, S., Fujii, K.: Computational Study of a Supersonic Base Flow Using LES/RANS Hybrid Methodology. AIAA Paper 2004-68 (2004)

13. Liou, M.S., Steffen, C.J.: A new flux splitting scheme. J. of Computational Physics 107, 23–39 (1993)
14. Meinke, M., et al.: A comparison of second- and sixth-order methods for large-eddy simulations. Computers & Fluids 31, 695–718 (2002)
15. Papp, J.L., Ghia, K.N.: Application of the RNG Turbulence Model to the Simulation of Axisymmetric Supersonic separated Base Flows. AIAA Paper 2001-027 (2001)
16. Renze, P., Meinke, M., Schröder, W.: LES of turbulent mixing in film cooling flows. To be published in Flow, Turbulence and Combustion (2007)
17. Renze, P., Meinke, M., Schröder, W.: LES of Film Cooling for Different Jet Fluids. J. of Aerospace Power 22(4), 521–530 (2007)
18. Rollstin, L.: Measurement of Inflight Base Pressure on an Artillery-Fired Projectile. AIAA Paper 87-2427 (1987)
19. Sagaut, P., et al.: Turbulent Inflow Conditions for Large-Eddy Simulation of Compressible Wall-Bounded Flows. AIAA Journal 42, 469–477 (2004)
20. Sahu, J., Heavey, K.R.: Numerical Investigation of Supersonic Base Flow with Base Bleed. J. of Spacecraft and Rockets 34, 61–69 (1997)
21. Sandberg, R.D., Fasel, H.F.: High Accuracy DNS of Supersonic Base Flows and Control of the Near Wake. In: Users Group Conference, DOD_UGC (2004)
22. Sandberg, R.D., Fasel, H.F.: Numerical investigation of transitional supersonic axisymmetric wakes. J. of Fluid Mechanics 563, 1–41 (2006)
23. Schlichting, H.: Grenzschicht-Theorie. In: Braun, G. (ed.), 5th edn. (1965)

Large Eddy Simulation of Nozzle Jet - External Flow Interaction

Y. Jin and R. Friedrich

Fachgebiet Strömungsmechanik
TU München
Boltzmannstr. 15, 85748 Garching, Germany
r.friedrich@lrz.tu-muenchen.de

Summary

Results of large eddy simulations (LES) of the interaction between nozzle and external flow on a sector of the full 3D computational domain are presented and compared with results of one- and two-equation turbulence models and with experiments. The LES results based on resolved scales turn out to predict mean flow quantities in closer agreement with experimental data than the turbulence models. Both prediction methods, however, underline the need to perform fully coupled simulations in order to capture the strong interaction between base and nozzle flow as reliably as possible.

1 Introduction

The study of the interaction between the rocket nozzle flow and the base flow is very important for the aero-thermodynamical design of future launchers. It could be even critical when reusable launchers have to be designed: The boundary layer separation induced by large exhaust plumes at the spacecraft upstream of the nozzle e.g. leads to a significant rise of the boat tail drag and to heavy thermal loads on the base structure. Boundary layer separation can also induce vibrations of the structure, which may be fatal to the stability and control effectiveness of the launcher.

Studies of such plume/base flow interactions by experimental and numerical methods can be traced back to the 1950s, while a large amount of literature appeared in the 1960s. Brewer and Craven [1] investigated the base flow field in a four engine clustered nozzle configuration. Their experiments revealed that a low energy resonance of known source was present in the base cavity. Later on, more experiments were performed to understand the coupling between nozzle and base flow, and to validate analytical and CFD results. Experimental studies of the rear flow field of AIRANE 5 were performed in France by Reijassse and Delery [12] in 1994. Henckels et al. [5] of the German Aerospace Centre (DLR) studied the interaction of the nozzle flow and the base flow in the DLR hypersonic wind tunnel H2K. The experiments showed that sting interactions may induce fluctuations of the plume shear layer of the order of 1 kHz in the upper part of the flow field.

Chapman and Korts [2] did the pioneering work of numerical simulation in 1954: they used an inviscid formulation for the overall flow and a viscous analysis for the

A. Gülhan (Ed.): RESPACE – Key Tech. for Reusable Space Sys., NNFM 98, pp. 57–81, 2008.
springerlink.com

mixing layer. Later on, several-analytical theories were developed with the aim of getting a deeper insight into the flow physics [16, 11]. With the progress in high performance super-computing, CFD became a valuable tool to study the interaction of the nozzle plume and the base flow. E. M. Houtman et al [6] computed the flow around a blunted cone-cylinder rocket model with a centrally protruding nozzle with several turbulence models, such as the Spalart-Allmaras model, the BSL model, the SST model and the $k-\omega$ model. The results show good agreement in those parts of the flow which are not dominated by viscous effects. But prediction of the separated base flow remains a very difficult problem. There are some other works which simulate this problem by CFD [10, 13, 3]. Nevertheless, the complex 3D base/nozzle flow field with a separated flow region and its interactions remain a challenging task to CFD.

Most of the previous work in this field has been done using the Reynolds averaged Navier-Stokes equations (RANS). In the meantime large-eddy simulation (LES) has become a promising tool to understand the space/time structure of turbulent flow fields. Moreover, it is not as expensive as DNS. Studies of the interaction of base and nozzle flow by LES, are still new in this field. The present research is funded by the RESPACE project, which consists in the cooperation of three university partners (RWTH Aachen, Uni. Stuttgart, and TU-Munich) and the DLR as coordinator, the work concentrates on "Key Technologies for Reusable Space Systems". In this contribution, Chapter 2 discusses the mathematical models. The results are shown in Chapter 3, including the unsteady interaction of nozzle/base flow and the comparison between RANS and LES results. Chapter 4 provides conclusions.

2 Physical and Mathematical Models

2.1 Governing Equations

In 3D Cartesian coordinates (x, y, z), the unsteady compressible Navier-Stokes equations can be expressed in conservative form as

$$\frac{\partial}{\partial t}(W)+\frac{\partial}{\partial x}(f-f_v)+\frac{\partial}{\partial y}(g-g_v)+\frac{\partial}{\partial z}(h-h_v)=0$$

$$W=\begin{pmatrix}\rho\\ \rho u\\ \rho v\\ \rho w\\ \rho E\end{pmatrix},\quad f=\begin{pmatrix}\rho u\\ \rho uu+p\\ \rho vu\\ \rho wu\\ \rho uH\end{pmatrix},\quad f_v=\begin{pmatrix}0\\ \tau_{xx}\\ \tau_{yx}\\ \tau_{zx}\\ \tau_{xx}u+\tau_{xy}v+\tau_{xz}w+k\dfrac{\partial T}{\partial x}\end{pmatrix},\quad g=\begin{pmatrix}\rho v\\ \rho uv\\ \rho vv+p\\ \rho wv\\ \rho vH\end{pmatrix} \tag{1}$$

$$g_v=\begin{pmatrix}0\\ \tau_{xy}\\ \tau_{yy}\\ \tau_{zy}\\ \tau_{yx}u+\tau_{yy}v+\tau_{yz}w+k\dfrac{\partial T}{\partial y}\end{pmatrix},\quad h=\begin{pmatrix}\rho w\\ \rho uw\\ \rho vw\\ \rho ww+p\\ \rho wH\end{pmatrix},\quad h_v=\begin{pmatrix}0\\ \tau_{xz}\\ \tau_{yz}\\ \tau_{zz}\\ \tau_{zx}u+\tau_{zy}v+\tau_{zz}w+k\dfrac{\partial T}{\partial z}\end{pmatrix}$$

where ρ is the density, u, v and w are the Cartesian velocity components, p is the pressure and E is the total energy. The shear stress tensor τ_{ij} is given by

$$
\begin{aligned}
\tau_{xx} &= \frac{2}{3}\mu\left(2\frac{\partial u}{\partial x} - \frac{\partial v}{\partial y} - \frac{\partial w}{\partial z}\right) \\
\tau_{yy} &= \frac{2}{3}\mu\left(2\frac{\partial v}{\partial y} - \frac{\partial u}{\partial x} - \frac{\partial w}{\partial z}\right) \\
\tau_{zz} &= \frac{2}{3}\mu\left(2\frac{\partial w}{\partial z} - \frac{\partial u}{\partial x} - \frac{\partial w}{\partial z}\right) \\
\tau_{xy} &= \mu\left(\frac{\partial u}{\partial y} + \frac{\partial v}{\partial x}\right) \\
\tau_{yz} &= \mu\left(\frac{\partial v}{\partial z} + \frac{\partial w}{\partial y}\right) \\
\tau_{xz} &= \mu\left(\frac{\partial u}{\partial z} + \frac{\partial w}{\partial x}\right)
\end{aligned}
\tag{2}
$$

where μ is the dynamic viscosity.

2.2 Turbulence Models

In the present paper, large eddy simulations (LES) as well as Reynolds averaged Navier-Stokes simulations (RANS) are performed to predict the impact of turbulence on the nozzle/base flow interaction.

Large eddy simulation

LES is based on the spatially filtered equations of motion. Any flow variable ϕ is decomposed into a large scale component or resolved field $\overline{\phi}^L$ and into a small scale or subgrid component

$$\phi = \overline{\phi}^L + \phi''^L \tag{3}$$

where the filter operation is defined by the convolution integral over the domain $\Omega = O(\Delta^3)$:

$$\overline{\phi}^L(x) = \int_\Omega G(x - x^*, \Delta)\phi(x^*)d^3x^* \tag{4}$$

where G is the filter function depending on the filter width Δ. The filter operation eliminates the high-wavenumber Fourier components of the flow variable ϕ. Common filters are the Gaussian filter, the top hat or box filter and the sharp cut-off or ideal filter.

Filtering the Navier-Stokes equations will introduce new unknown quantities which have to be modelled in order to close the system. As an example we write down the subgrid scale stress tensor

$$\tau_{ij}^{sgs} = -\left(\overline{\rho u_i u_j}^L - \overline{\rho}^L \overline{u}_i^L \overline{u}_j^L\right) \tag{5}$$

The eddy viscosity assumption or Boussinesq hypothesis is used to model this tensor. For consistency, the deviatoric part of τ_{ij}^{sgs} is related to the large scale strain rate tensor as:

$$\tau_{ij}^{sgs} - \frac{1}{3}\tau_{kk}^{sgs}\delta_{ij} = \mu_t^{sgs}\left(\frac{\partial \overline{u}_i^L}{\partial x_j} + \frac{\partial \overline{u}_j^L}{\partial x_i} - \frac{2}{3}\frac{\partial \overline{u}_k^L}{\partial x_k}\delta_{ij}\right) \tag{6}$$

where τ_{kk}^{sgs} represents the subgrid scale kinetic energy. The sgs-eddy-viscosity μ_t^{sgs} is then expressed by the Smagorinsky's subgrid scale model [14]. In a similar manner sgs-heat fluxes are modelled.

Statistical models

The Reynolds averaged Navier-Stokes equations are used in RANS computations, in which the Reynolds stresses, turbulent heat fluxes and some other terms must be modelled. In the present computations an eddy viscosity assumption is used to model the Reynolds stresses as follows

$$-\overline{\rho u_i' u_j'} = \mu_t\left(\frac{\partial \tilde{u}_i}{\partial x_j} + \frac{\partial \tilde{u}_j}{\partial x_i}\right) - \frac{2}{3}\delta_{ij}\left(\mu_t \frac{\partial \tilde{u}_k}{\partial x_k} + \overline{\rho} k\right) \tag{7}$$

The eddy viscosity μ_t is not a fluid property, but a property that depends on the local turbulence structure. The variable k represents the turbulent kinetic energy, defined as $k = 1/2\left(\overline{\rho u_i' u_i'}/\overline{\rho}\right)$. In Eq. (7) the tilde denotes a mass-weighted average and the dash the corresponding fluctuation (deviation from the instantaneous variable). In the present paper, results from the 1-equation Spalart-Allmaras model and from the 2-equation $k-\omega$ model are compared with LES results.

Spalart-Allmaras model

Using empiricism and arguments of dimensional analysis, Galilean invariance, and selective dependence on the molecular viscosity, the transport equation for the turbulent viscosity is assembled in the Spalart-Allmaras model [15]:

$$\frac{D\tilde{\gamma}}{Dt} = G_\gamma + \frac{1}{\sigma}\left[\nabla \cdot ((\gamma + \tilde{\gamma})\nabla\tilde{\gamma}) + C_{b2}(\nabla\tilde{\gamma})^2\right] - Y_\gamma \tag{8}$$

The eddy viscosity is defined as:

$$\mu_t = \overline{\rho}\tilde{\gamma} f_{v1} \tag{9}$$

In the buffer layer and viscous sublayer, the damping function f_{v1} is defined as:

$$f_{v1} = \frac{\lambda^3}{\lambda^3 + C_{v1}^3} \tag{10}$$

λ is a function of the local variable

$$\lambda \equiv \frac{\tilde{\gamma}}{\nu} \tag{11}$$

where ν is the molecular kinematic viscosity. G_γ is the production of turbulent viscosity and Y_γ is the destruction of turbulent viscosity that occurs in the near-wall region due to wall blocking and viscous damping. σ and C_{b2} are constants.

Standard $k-\omega$ model

The standard $k-\omega$ model is an empirical model based on transport equations for the turbulence kinetic energy (k) and the specific dissipation rate (ω) [18]. The equations are as follows

$$\frac{\partial}{\partial t}(\bar{\rho} k) + \frac{\partial(\bar{\rho}\tilde{u}_i k)}{\partial x_i} = \frac{\partial}{\partial x_j}\left(\Gamma_k \frac{\partial k}{\partial x_j}\right) + G_k - Y_k + S_k \tag{12}$$

$$\frac{\partial}{\partial t}(\bar{\rho} \omega) + \frac{\partial(\bar{\rho}\tilde{u}_i \omega)}{\partial x_i} = \frac{\partial}{\partial x_j}\left(\Gamma_\omega \frac{\partial \omega}{\partial x_j}\right) + G_\omega - Y_\omega + S_\omega \tag{13}$$

In these equations, G_k represents the generation of turbulence kinetic energy due to mean velocity gradients. G_ω represents the generation of ω. Γ_k and Γ_ω describe the effective diffusivity of k and ω, respectively. Y_k and Y_ω are the dissipation of k and ω due to turbulence. S_k and S_ω are user-defined source terms. The turbulent viscosity, μ_t, is computed by combining k and ω as follows:

$$\mu_t = \alpha^* \frac{\bar{\rho} k}{\omega} \tag{14}$$

where α^* is the coefficient to damp the turbulent viscosity which causes a low-Reynolds-number correction.

2.3 Numerical Method

Equation (1) can be written in the following form

$$\frac{\partial \vec{W}}{\partial t} + \frac{\partial \vec{F}_i}{\partial x_i} = \vec{S} \tag{15}$$

Integration of Eq. (15) in cell (*i*, *j*, *k*) leads to a set of ordinary differential equations in time

$$\frac{d}{dt}\left(V_{i,j,k}W_{i,j,k}\right)+R_{i,j,k}=0 \tag{16}$$

where $R_{i,j,k}$ is the residual, consisting of the net flux out of cell (*i*, *j*, *k*).

For unsteady applications, a compact three-stage Runge-Kutta time stepping scheme which is of third order accuracy is used:

$$\begin{aligned} W^{(1)} &= W^{n} - \alpha_1 \frac{\Delta t}{V} R_0 \\ W^{(2)} &= W^{(1)} - \alpha_2 \frac{\Delta t}{V} R_1 \\ W^{(3)} &= W^{(2)} - \alpha_3 \frac{\Delta t}{V} R_2 \end{aligned} \tag{17}$$

where

$$\begin{aligned} R_0 &= R\left(W^{n}\right) \\ R_1 &= R\left(W^{(1)}\right)+\xi_1 R_0 \\ R_2 &= R\left(W^{(2)}\right)+\xi_2 R_1 \end{aligned} \tag{18}$$

In Eqs. (17, 18) the cell indices have been neglected for simplicity. The values proposed by Lowery and Reynolds [19] for the above coefficients are:

$$\begin{aligned} \alpha_1 &= 1/2 \\ \alpha_2 &= 0.9106836025229591 \\ \alpha_3 &= 0.3660254037844387 \\ \xi_1 &= -0.6830127018922193 \\ \xi_2 &= -4/3 \end{aligned} \tag{19}$$

Results from direct numerical simulations show that the subgrid scale stresses are of the order $O(h^2)$. Therefore, a large-eddy simulation has to be conducted with a scheme of at least second order accuracy. Additionally, the numerical dissipation of the approximation has to be as low as possible, since the small scales of the turbulent flow are significantly affected by numerical dissipation.

For this reason, central difference schemes are very popular in LES of incompressible flows. Unfortunately, the standard central difference approximations tend to be unstable in compressible flows, unless some kind of artificial numerical dissipation is added. To overcome these difficulties we use a mixed central-upwind AUSM scheme for the discretization of the inviscid fluxes that is based on the method of Liou and Steffen [8]. Reference [9] compares the second-order AUSM and sixth-order compact schemes by simulating planar turbulent jets. It is stated that the two schemes show hardly any difference in the quality of the solutions. The second-order AUSM scheme, however, is computationally more efficient. The AUSM scheme is described below, using instantaneous variables without referring to any spatial filtering or statistical averaging.

The inviscid fluxes of Eq. (15) can be divided into a convective and a pressure term

$$\vec{F}_i = \vec{F}_i^c + \vec{F}_i^p = \frac{u_i}{c}\vec{f}_i^c + \vec{F}_i^p = \frac{u_i}{c}\begin{pmatrix} \rho c \\ \rho c u_j \\ \rho c(E + p/\rho) \end{pmatrix} + \begin{pmatrix} 0 \\ \delta_{ij} p \\ 0 \end{pmatrix} \tag{20}$$

The numerical flux on the cell face, e.g. $i \pm 1/2$, j, k, is computed by

$$F_i^c = \frac{1}{2}\left[\frac{Ma_i^+ + Ma_i^-}{2}\left(f_i^{c+} + f_i^{c-}\right) + \frac{\left|Ma_i^+ + Ma_i^-\right|}{2}\left(f_i^{c+} - f_i^{c-}\right)\right]_{i\pm1/2,j,k} \tag{21}$$

where the fluxes $f_i^{c\pm}$ and the Mach numbers $Ma_i^{\pm}$ are determined by left and right interpolated variables obtained using a monotonic upstream centered scheme in a MUSCL approach [7].

The pressure term can be computed according to Ref. 8 as follows

$$p^{\pm} = p^{\pm}\left(\frac{1}{2} \pm \kappa Ma_i^{\pm}\right) \tag{22}$$

The parameter κ, which defines what kind of pressure splitting is used and thus determines the numerical dissipation of the scheme, ranges from 0 to 1/96.

3 Results and Discussion

3.1 Mesh and Aerodynamic Parameters

The structure of the rocket nozzle is shown in Fig. 1. As presented in the picture, a settling chamber and an integrated honeycomb insert are used to reduce the turbulence generated by the manifold of the air supply. So, at the inlet of the nozzle, a random homogeneous perturbation of low amplitude is to be expected. The total pressure and total temperature at the nozzle inlet are 20 bar and 900 K. The task of TUM in this cooperation was to simulate the flow inside the nozzle and downstream of the nozzle exit plane. To achieve this, flow conditions outside the nozzle must be known, because the nozzle jet does not develop independently of the external flow. Indeed there is a strong interaction between the jet flow and the base flow and it is of great interest and importance to understand the characteristics of this interaction. Therefore, we extend our computational domain to the external region and incorporate the base flow. While the free-stream Mach number of the chosen flow case is 5.3, the Mach number behind the front bow shock (outside the boundary layer of the model) is 5.12. This value is obtained from simulations at RWTH Aachen and used as a boundary condition for our computation. The free-stream total temperature $T_{0\infty}$ is 600 K. The free-stream total pressure $p_{0\infty}$ is 3 bar. The Reynolds number based on the free-stream velocity is 2.6×10^{6} m^{-1}. The wall temperature of the inner side of the settling chamber and the whole nozzle is 320 K. It is also the wall temperature of the external surface, comprising nozzle, base shoulder and cylindrical body.

Since the rocket downstream of the sting is axisymmetric, it is possible to compute only a sector of 60 degrees in the LES. This is what we did, but, as an alternative, the complete domain (involving 360 degrees in circumferential direction) will also be computed in the future. The global mesh can be seen in Fig. 2. $1/6^{th}$ of the domain has 1,305,000 cells, while the whole domain has 8,924,400 cells. The small domain is divided into 9 blocks and the whole domain comprises 54 blocks for MPI parallel

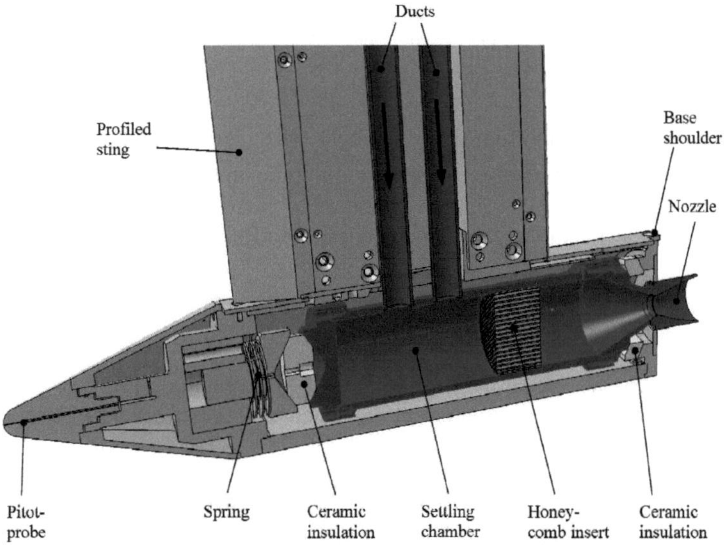

Fig. 1. Structure of the rocket nozzle

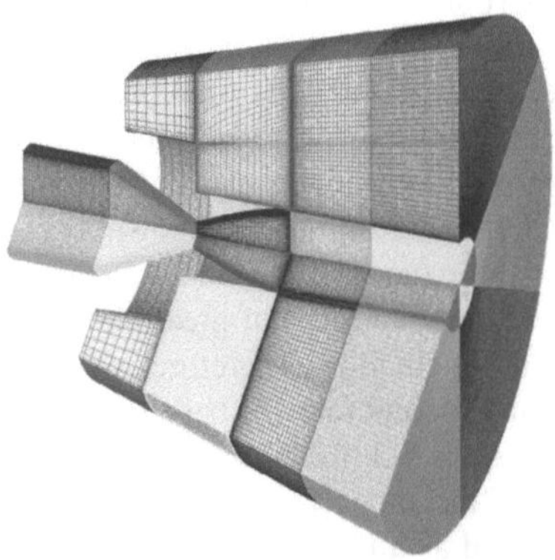

Fig. 2. Global mesh

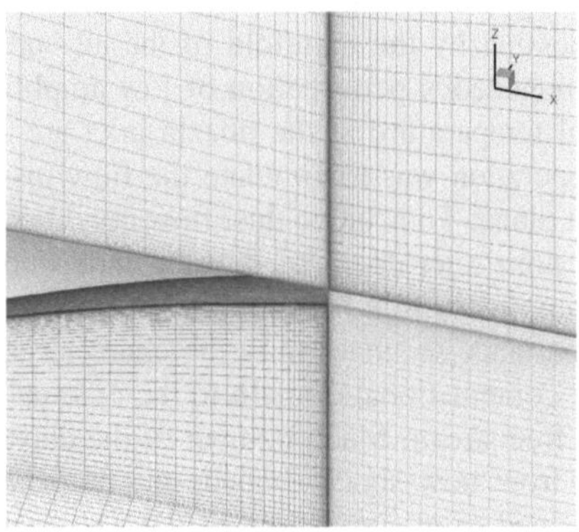

Fig. 3. Zoom into mesh near the nozzle exit

computation. Figure 3 shows a zoom into the mesh near the exit of the nozzle, where we note the higher mesh resolution near the wall.

In order to estimate the mesh size for the LES along the cylindrical rocket body, we use the skin-friction coefficient as obtained from Prandtl's law [17] for a turbulent flat plate boundary layer:

$$C_f \approx 0.0592\, Re_x^{-1/5} \tag{23}$$

The friction velocity u_τ, being the square root of the wall shear stress divided by the wall density, is:

$$u_\tau \equiv \sqrt{\frac{C_f}{2}} \cdot U_\infty \tag{24}$$

In Eq. (24) the ratio of wall to free-stream density is assumed of order 1. The wall-normal size of the first cell adjacent to the external wall can then be estimated to

$$l = \Delta z(1) = \frac{\mu_\infty}{\rho_\infty u_\tau} \approx 7.55 \times 10^{-6}\, m \tag{25}$$

where μ_∞ is the free stream viscosity, ρ_∞ is the free stream density.

The estimation of the mesh size normal to the nozzle wall is based on the assumption that a laminar boundary layer develops downstream of the honeycomb, and is accelerated up to the throat and beyond. For a compressible laminar boundary layer F.M. White [17] provides the following wall shear stress estimate

$$\tau_w \approx 0.664\, Re_x^{-0.5} \frac{\rho_\infty}{2} U_\infty^2 \sqrt{C_w}, \quad C_w = \rho_w \mu_w / (\rho_\infty \mu_\infty) \tag{26}$$

The Chapman-Rubesin parameter at the wall, C_w, is assumed $O(1)$. The Eq. (26) is applied at the position of the throat, using the critical values for the density and velocity as free-stream values. Then the wall-normal size of the first cell in the nozzle is approximately

$$l = \frac{\mu_\infty}{\sqrt{\rho_\infty \tau_w}} \approx 0.24 \times 10^{-6}\, m \tag{27}$$

The reference values to normalize the numerical results are as follows

$$\begin{aligned}
p_{ref} &= p_{0\infty} = 3\, bar \\
T_{ref} &= T_{0\infty} = 600\, K \\
\rho_{ref} &= \frac{p_{ref}}{RT_{ref}} = 1.742 kg/m^3 \\
V_{ref} &= a_\infty = \sqrt{\gamma R T_{ref}} = 491\, m/s \\
l_{ref} &= 0.108\, m
\end{aligned} \tag{28}$$

3.2 Instantaneous Flow Field

All results shown in this and the following sections are obtained on 1/6th of the global mesh. Fig. 4 compares the large eddy simulation results with the experimental data of Henckels and Gruhn [5]. The free stream Mach number is 5.3 in Fig. 4a and Fig. 4b, however, there is no external flow in Fig. 4c and Fig. 4d. The contours of the temperature computed by LES clearly show the position of the plume shear layer, which agrees very well with that obtained in the experiment. Compared to the results with free-stream velocity, the plume shear layer without external flow expands to a

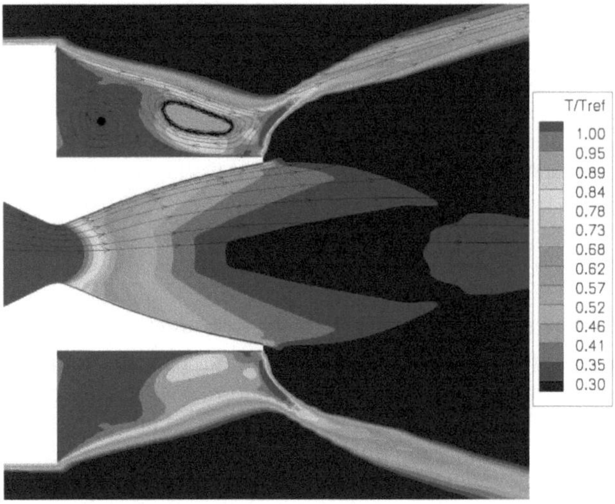

a) *LES,* $p_{nozzle}/Pa = 34.4$*, Ma = 5.3*

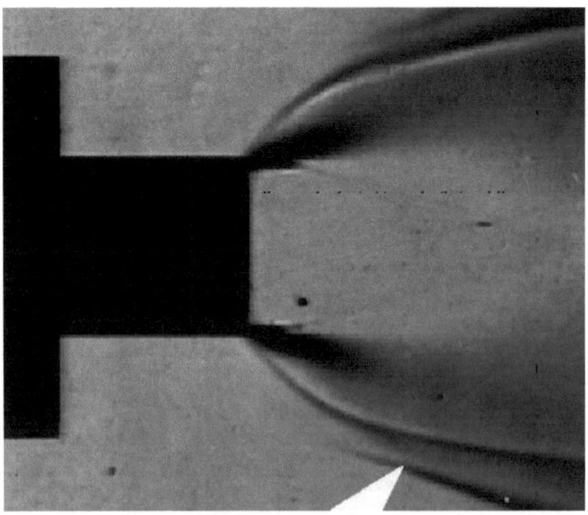

b) *Exp.* $p_{nozzle}/Pa = 36.2$*, Ma = 5.3, Henckels, Gruhn & Gülhan*

Fig. 4. Comparison of the LES results (temperature contours) with the experiment of [5]

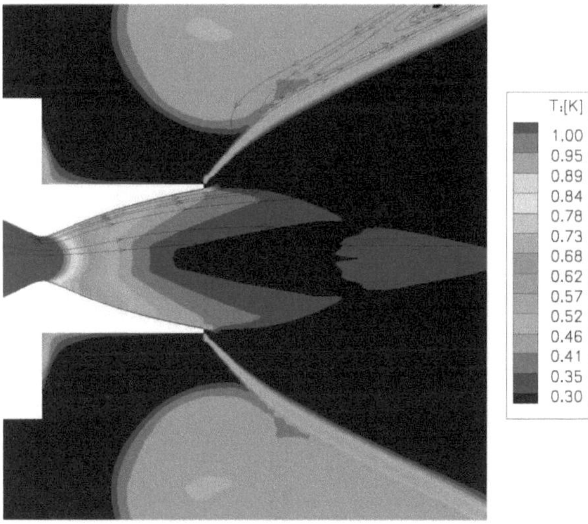

c) *LES*, $p_{nozzle}/Pa = 34.4$, $Ma = 0$

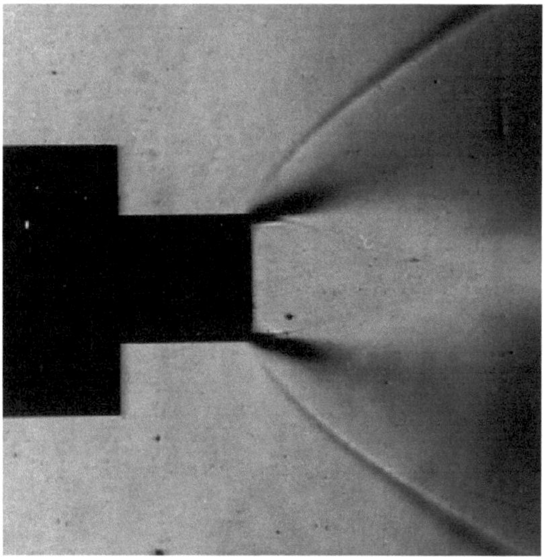

d) *Exp.* $p_{nozzle}/Pa = 36.2$, $Ma = 0$, *Henckels, Gruhn & Gülhan*

Fig. 4. (*continued*)

wider angle. Figure 4a and Fig. 4c also contain the instantaneous streamlines. A large recirculation zone can be seen in the base region for $Ma_\infty = 5.3$, which starts to interact with the plume shear layer near the exit of the nozzle. An especially high temperature region is found in that early interaction zone.

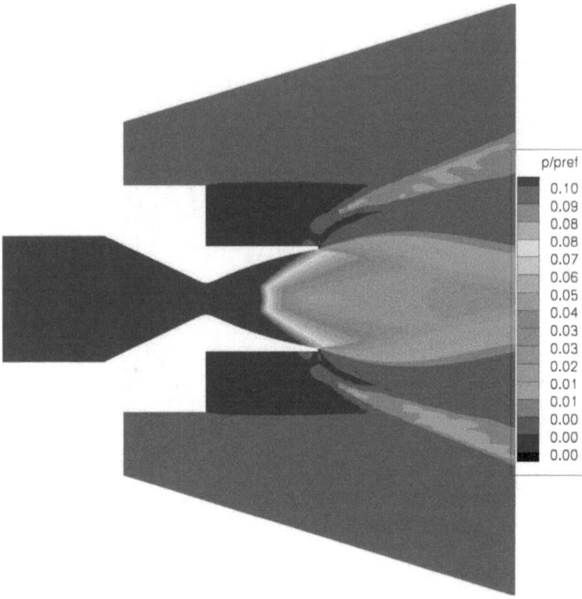

a) *pressure*

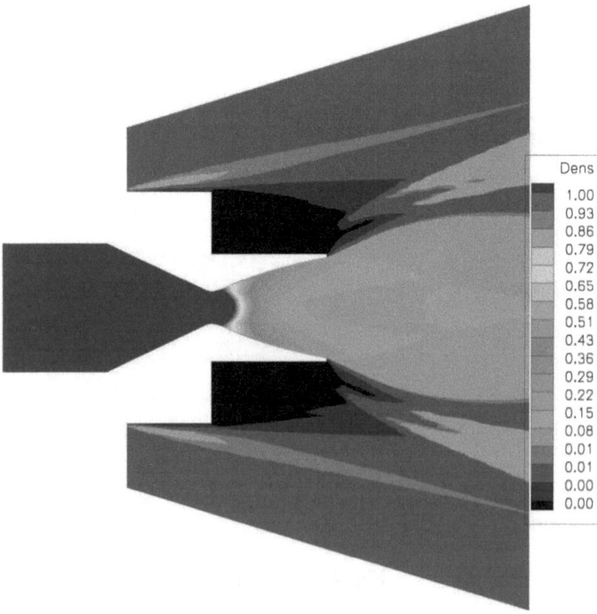

b) *density*

Fig. 5. Instantaneous flow field

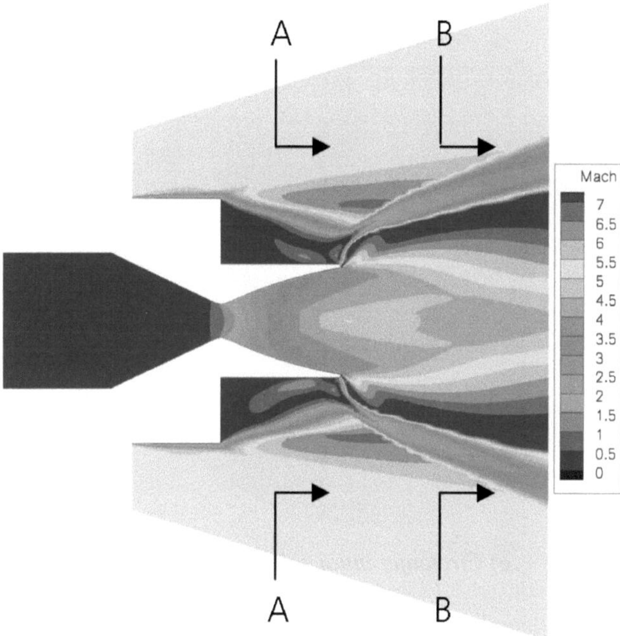

c) *Mach number*

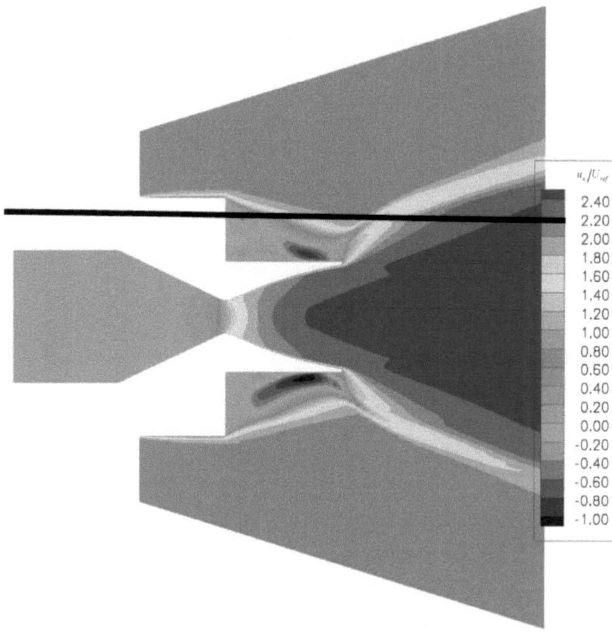

d) *axial velocity*

Fig. 5. (*continued*)

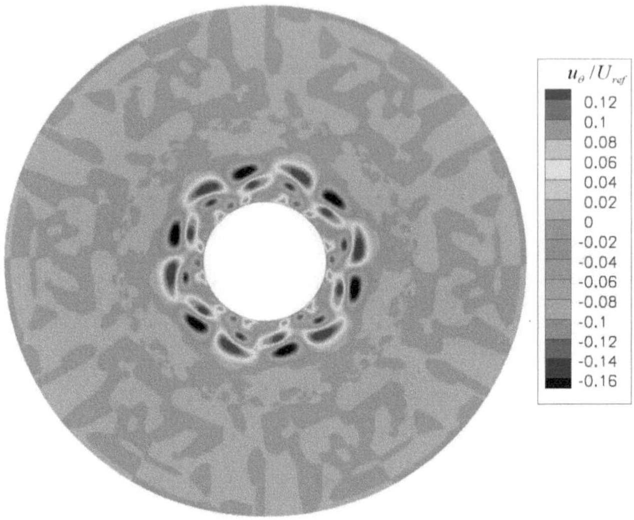

e) *Circumferential velocity, Sec. A-A*

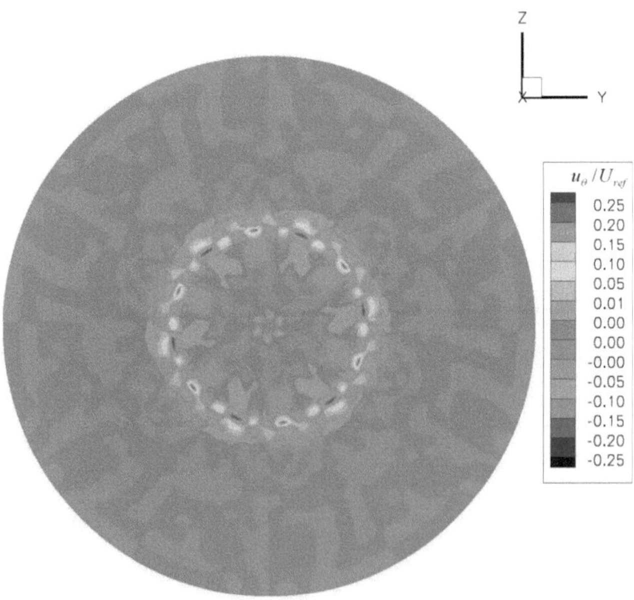

Fig. 5. (*continued*)

Figure 5 shows the contours of the other normalized variables: pressure, density, Mach number and axial velocity. In Fig. 5c and Fig. 5d, another shear layer starting at the base shoulder and merging with the plume shear layer can be seen. It encloses the recirculating flow. The Mach number contours in Fig. 5c also reveal that the nozzle flow reaches Mach numbers around 7 due to expansion effects of the jet flow downstream of the nozzle exit. Figure 5e and Fig. 5f show the circumferential velocity in the base region (section A-A) and in the plume shear layer (Section B-B) .It can be seen that there are large scale fluctuations in the base flow and they break into small scale fluctuations near the wall (Fig. 5e). A large-scale fluctuation can also be seen in the plume shear layer (Fig. 5f). In order to observe the interaction of the base flow and the plume shear layer, Fig. 6 shows the evolution of the normalized circumferential velocity U_θ and temperature T at the line where $\theta = \pi / 6,\ r = 0.38$. Point "A" in Fig. 6 indicates the position of the shear layer. The base region is to the left of "A".

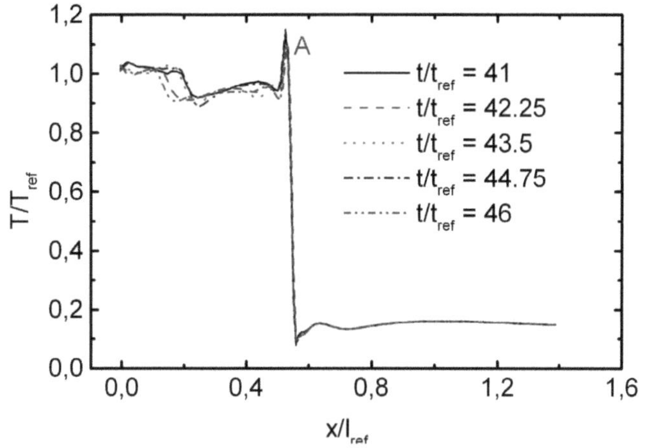

a) *Temperature*

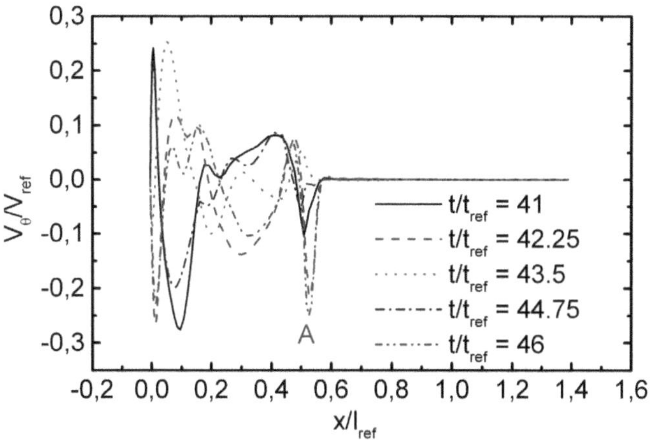

b) *Circumferential velocity*

Fig. 6. Time dependentent variables of the base flow and of the plume shear layer, $\theta = \pi / 6,\ r / l_{ref} = 0.38$

Figure 7 presents the instantaneous fields of temperature T and circumferential velocity V_θ. From the snapshots, it can be seen that large scale velocity fluctuations are generated in the shear layer of the base region, entraining high temperature

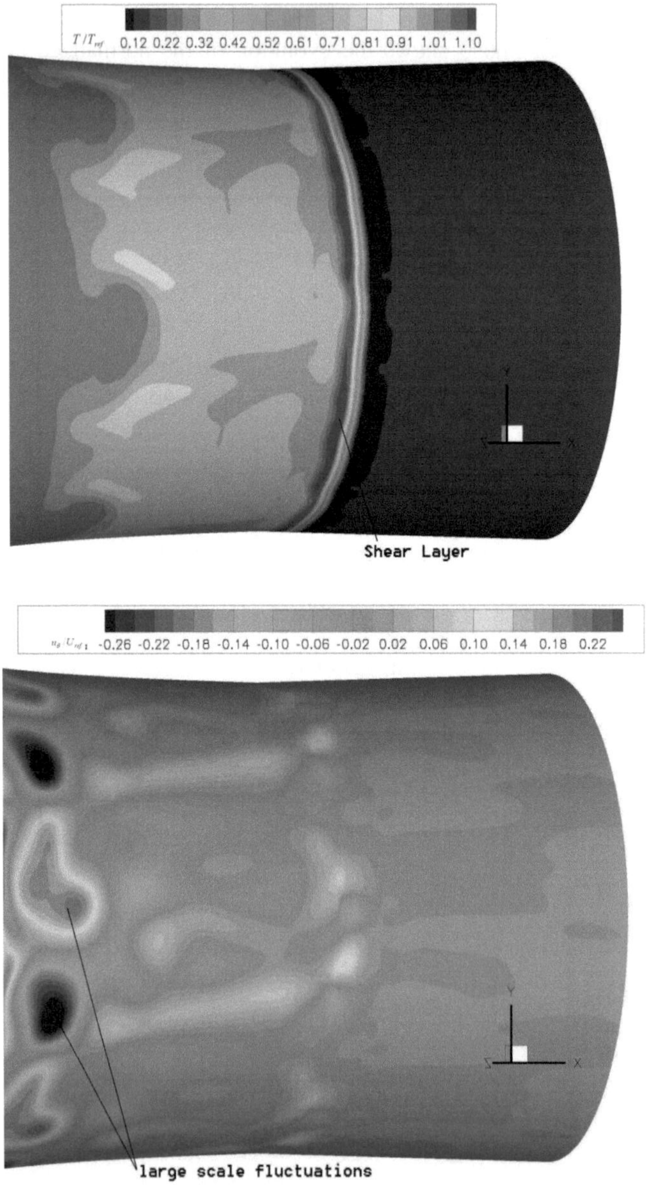

a) $t/t_{ref} = 41$

Fig. 7. Instantaneous flowfields which indicate the interaction of the shear layer and the base flow

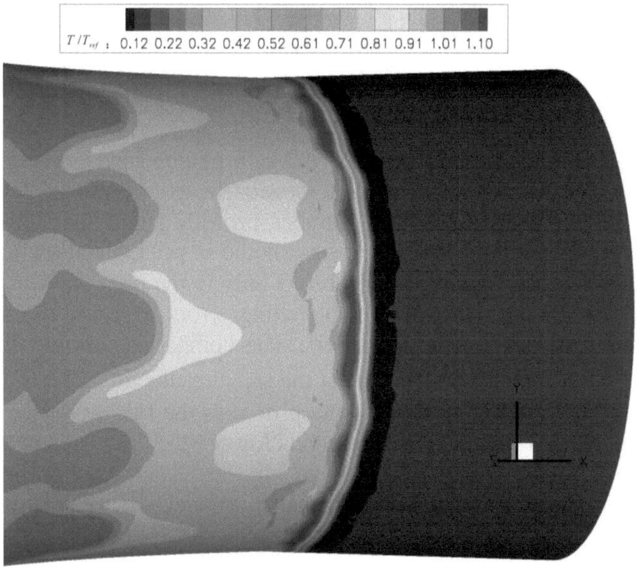

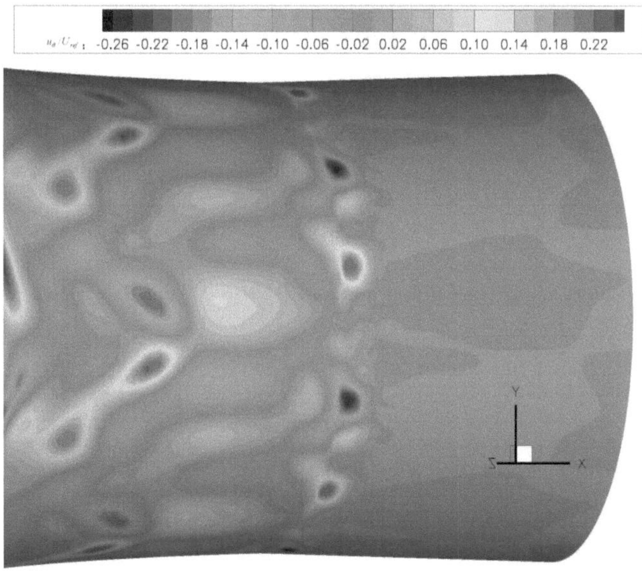

b) $t/t_{ref} = 46$

Fig. 7. (*continued*)

fluctuations. These fluctuations spread downstream in the course of time and break into smaller scale fluctuations.

3.3 Nozzle Flow

Figures 8a and 8b contain radial profiles of Mach number and pressure coefficient obtained from LES and RANS results at the nozzle exit. Profiles of the pressure coefficient along the wall and Mach number profiles along the nozzle axis are presented in Figs. 8c and 8d. The LES results should in principle allow for validation of RANS results because LES modelling refers only to the high-wavenumber part of the turbulence energy spectrum. Hence modelling errors have lower impact on the final result than statistical modelling errors. The comparison shows that Wilcox's two-equation $k-\omega$ model predicts almost the same results as the LES. The Spalart-Allmaras' model provides the same pressure profile as the LES at the wall and at the nozzle exit (Fig. 8b and Fig. 8c), but it predicts lower Mach numbers near the axis in the divergent part of the nozzle (Fig. 8a and Fig. 8d). The fact that the pressure is relatively insensitive to changes in a mathematical model is a well-known feature in aerodynamics.

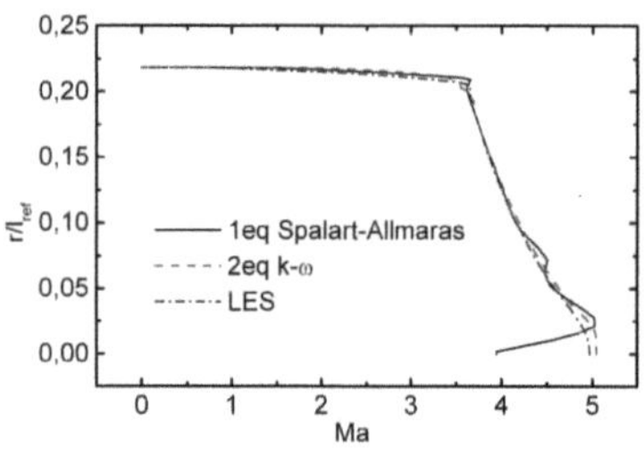

a) *Mach number profile at the nozzle wall*

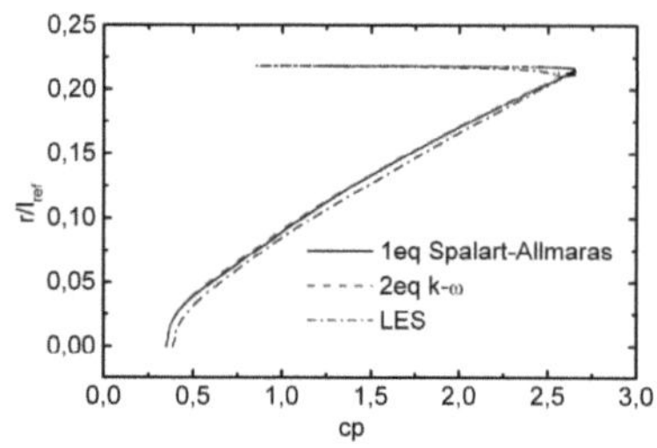

b) *Mach number profile at the axis*

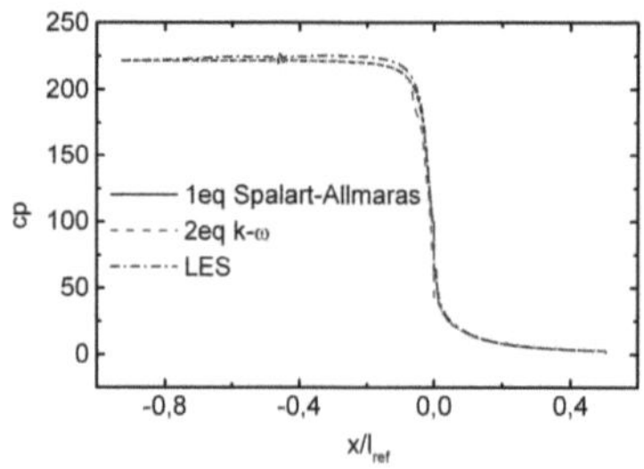

c) *Pressure coefficient profile at the nozzle wall*

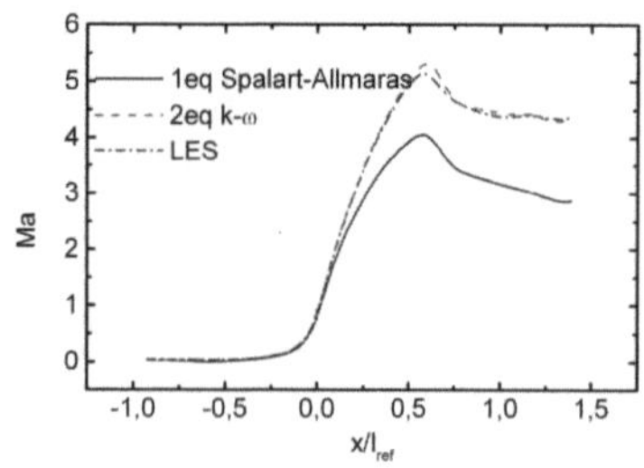

d) *Mach number profile at the axis*

Fig. 8. Numerical results of the nozzle flow

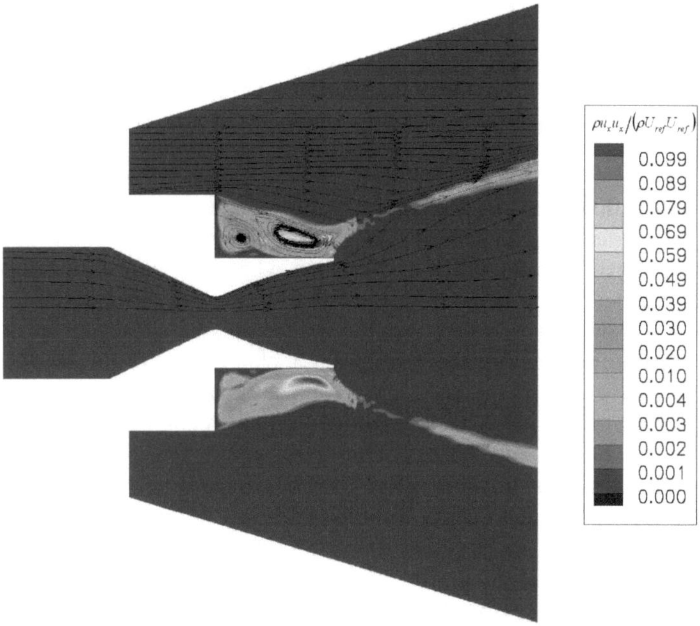

a) *Normal Reynolds stress,* $\overline{\rho u_x u_x} / \overline{\rho} u_{ref} u_{ref}$

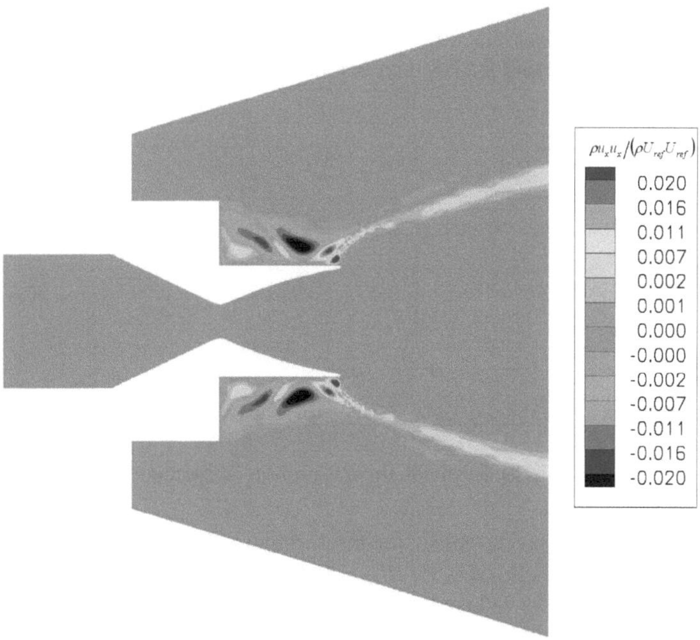

b) *Shear Reynolds stress,* $\overline{\rho u_x u_r} / \overline{\rho} u_{ref} u_{ref}$

Fig. 9. Contours of Reynolds stresses

3.4 Base Flow and Plume Shear Layer

As we know from the instantaneous flow field (Fig. 5 to Fig. 7), turbulence plays an important role in the base flow and the plume shear layer. Both start interacting near the exit of the nozzle. Figure 9 shows contours of the normal Reynolds stress ($\overline{\rho u_x u_x} / \bar{\rho} u_{ref} u_{ref}$) and of the Reynolds shear stress ($\overline{\rho u_x u_r} / \bar{\rho} u_{ref} u_{ref}$). We note strong turbulent flow in the base region, which is characterized by large scale vortical motions and also in the plume shear layer. The turbulence in the plume shear layer is amplified in streamwise direction. In the zone where the base flow and the plume shear layer interact, we observe the largest values of the Reynolds stresses.

Figure 10 shows profiles of the Reynolds stresses at different downstream positions. The solid lines are the profiles in a plane close to the nozzle exit. It is evident that the plume shear layer close to the nozzle exit produces the highest Reynolds stress amplitudes. As one moves downstream (x increases), the peak positions move outward and the peak amplitudes undergo a strong reduction before they increase again near $x/l_{ref} = 0.888$. This amplification is due to the fact that the plume shear layer there gets energy from the high speed flow region outside the base flow shear layer.

In order to demonstrate the performance of the k-ω-model, Mach number profiles, predicted by LES and RANS, are plotted in Fig. 11a, b. Although both results are computed on the same grid, there are some significant differences. It turns out that the statistical model underpredicts the strong overexpansion of the flow close to the nozzle edge which leads to Mach numbers above 7. The S-A model provides very similar and therefore unsatisfactory results. Further downstream ($x/l_{ref} = 0.888$) the agreement between LES and RANS data is much better.

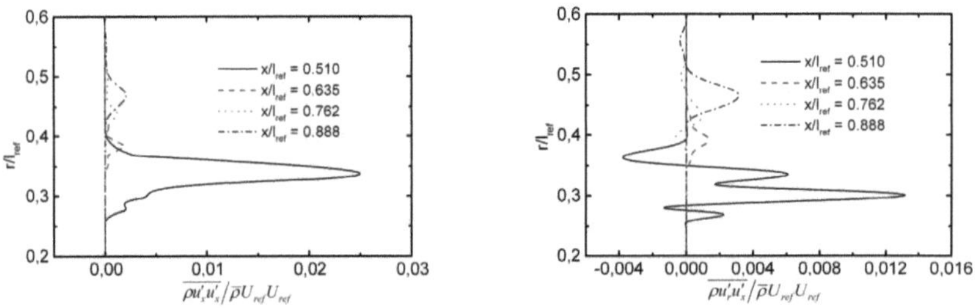

a) *Normal Reynolds stress* **b)** *Shear Reynolds stress*

Fig. 10. Reynolds stresses profiles in the wake

Of great interest is also the result for the turbulent kinetic energy in Figs. 11c, d. Close to the nozzle exit ($x/l_{ref} = 0.51$, $r/l_{ref} = 0.27$) where the Mach number peaks, the turbulent kinetic energy is reduced because of flow acceleration. In the plume shear layer itself, where the Mach number undergoes a strong reduction, the turbulent kinetic energy has its peak. This is a recently observed compressibility effect in mixing layers [4].

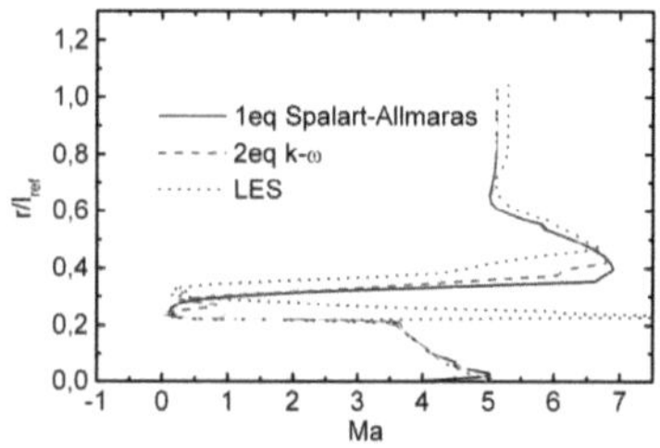

a) *Near the nozzle exit* ($x/l_{ref} = 0.51$, *section A-A*)

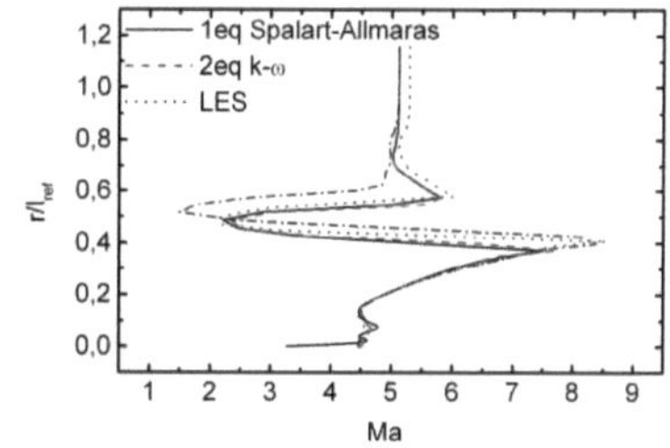

b) *In the wake* ($x/l_{ref} = 0.888$, *section B-B*)

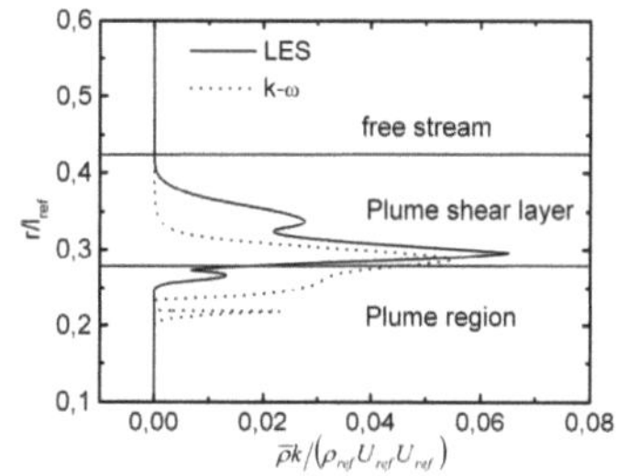

c) *Near the nozzle exit* ($x/l_{ref} = 0.51$, *section A-A*)

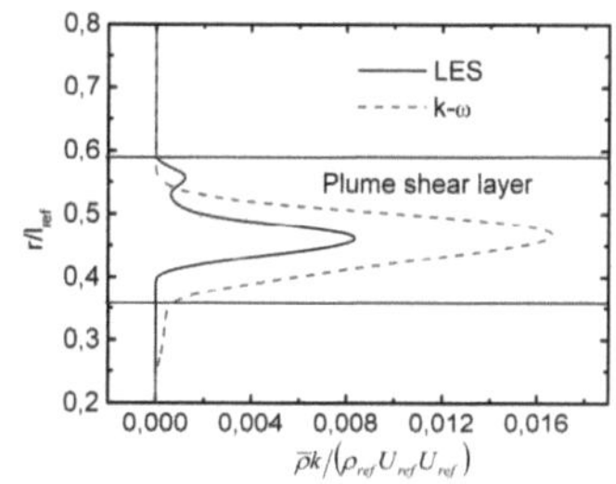

d) *In the wake* ($x/l_{ref} = 0.888$ *section B-B*)

Fig. 11. Mach number and turbulent kinetic energy profiles from LES and RANS computations near the nozzle exit and in the wake

3.5 Fluctuations of the Plume Shear Layer

Pressure coefficients were measured by scientists from DLR at 14 points on the surface of the model spacecraft, the positions of which are indicated in Fig. 12. Figure 13 shows the history of the pressure coefficients at points "5" and "7". The large eddy simulation predicts a little lower pressure than the experiment. But considering the uncertainty of the wall temperature in the base region, large eddy simulation provides a satisfactory prediction.

According to Ref. 6, RANS computations always predict lower pressure than the experiment in the base region. This is so, because it is a very difficult task to simulate the base flow. In the present work, both the SA and the two equation $k-\omega$ model predict even lower pressure coefficients than the LES (Fig. 14). This is another result underlining the power of large-eddy simulation.

The signals of pressure coefficients are transferred to Fourier space in Fig. 15. The horizontal coordinates represent the reduced frequency *k*, which is computed from

$$k = 2\pi f \frac{l_{ref}}{V_{ref}} \tag{29}$$

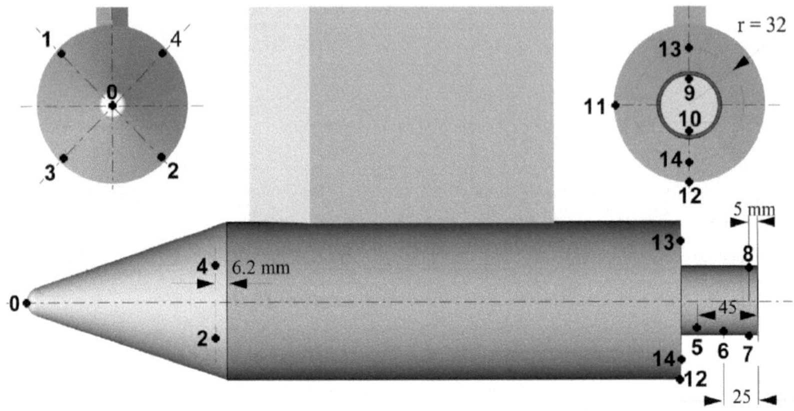

Fig. 12. Location of the pressure orifices

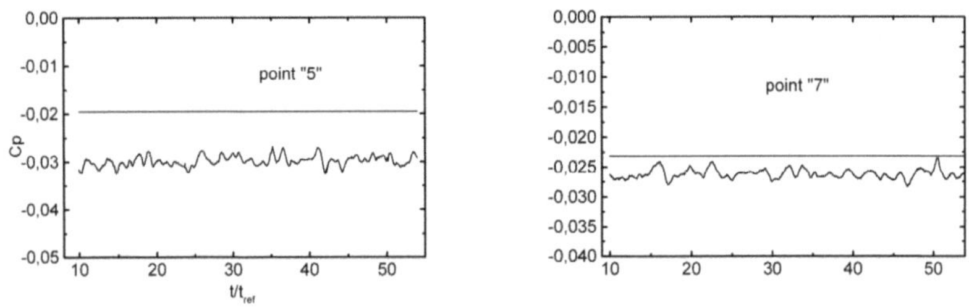

a) *Point "5"*

b) *Point "7"*

Fig. 13. History of the pressure coefficients of points in the base region versus experiment

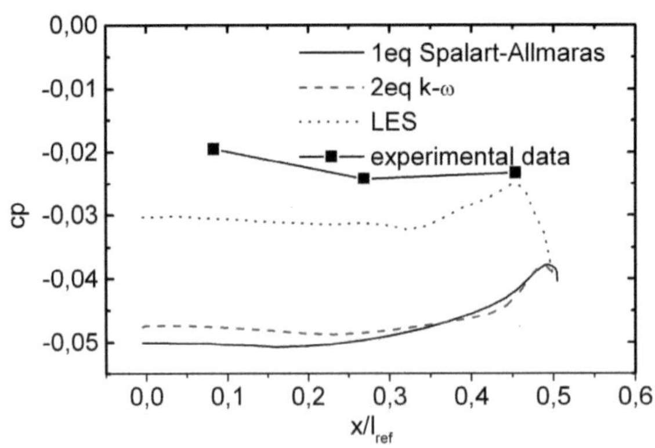

Fig. 14. Comparison of the pressure coefficients predicted by RANS, LES and the experiments in the base region

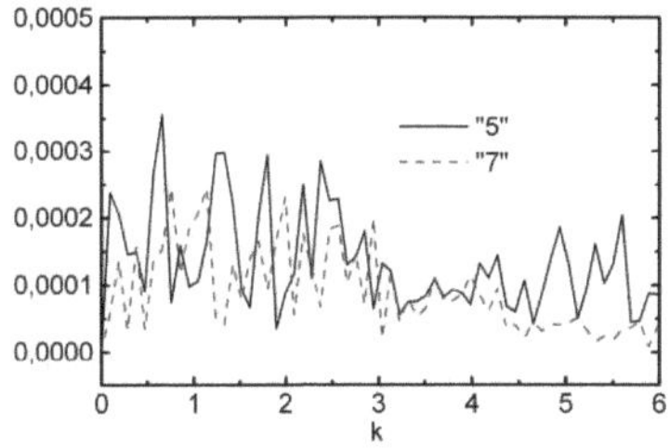

a) *Frequency spectrum in the base region*

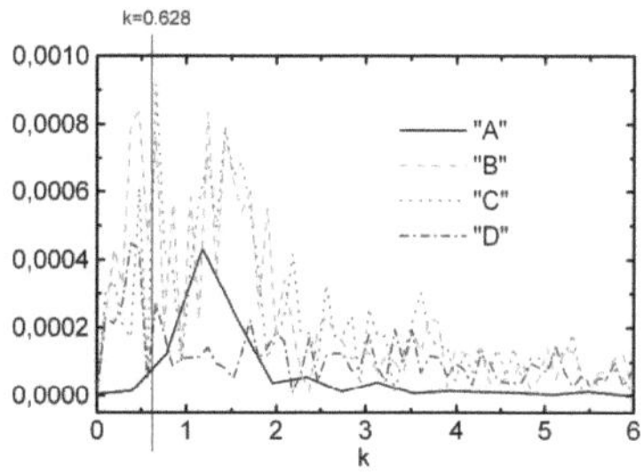

b) *Frequency spectrum in the plume shear layer*

Fig. 15. Frequency spectrum of the fluctuations in the base region and the plume shear layer

The vertical coordinates are the amplitudes of the pressure coefficients at the reduced frequency k. The Fourier spectra of Cp at point “5” and “7” (see Fig. 15a) show the nonlinear fluctuation in the base region, the frequency of which has a wide range of values (from 0 ~ 6). Figure 15b shows the frequency spectrum inside the plume (point “A”) and near the plume shear layer (points “B”, “C”, “D”). The positions of these points are shown in Fig. 9b. The spectrum shows that there is a unique frequency ($k = 1.25$, $f = 2000$ Hz) in the plume near the exit of the nozzle (point “A”). Where the plume flow is coupled with the base flow near the exit of the nozzle, the fluctuating energy at points “B” and “C” peaks at reduced frequencies ranging from 0.5 to 1.8 (800 Hz ~ 2086 Hz). The higher frequency in the base region (1.8 ~ 6, Fig. 10a) gradually decays in the streamwise direction. When observing a point that is far away from the base region, only a low value of the reduced frequency (about 0.5, 800 Hz) can be found (point “D”) with maximum fluctuating energy. The frequency spectrum confirms the phenomenon observed in the experiment of DLR, which reports a distinct frequency of the order of about 1 kHz.

4 Conclusions

The base flow/plume shear layer interaction is studied by large eddy simulation on $1/6^{th}$ of the full global mesh. The position of the plume shear layer predicted by large eddy simulation compares very well with the experimental results. Large eddy simulation predicts turbulent flow in the base region and near the plume shear layer. The instantaneous flow field shows that large-scale turbulent structures originate near the shear layer surrounding the recirculating flow, and break into small scale fluctuations further downstream where the two shear layers start interacting.

The LES results are used to validate the S-A and the two-equation $k-\omega$ model. The comparison shows that the $k-\omega$ model predicts results close to those of the LES in the nozzle region. But the S-A model predicts too low Mach numbers along the nozzle axis downstream of the nozzle exit.

The distribution of the Reynolds stresses shows that the turbulence is especially strong near the interface of the base flow and the plume shear layer. Here both the

S-A and the $k-\omega$ model predict Mach number profiles which differ from those of the LES results. The RANS data agree better with LES data in zones which are far away from the base region.

LES predicts somewhat lower pressure coefficients than the experiment, but still in closer agreement with the experiment than the RANS results. The frequency spectrum of *Cp* in the base flow region has a wide range of values, which has a clear influence on the plume shear layer near the exit of the nozzle. The frequency spectrum also confirms that the plume shear layer fluctuates with a distinct frequency of about 1 kHz.

References

1. Brewer, E., Craven, C.: Experimental investigation of base flow field at high altitude for a four-engine clustered nozzle configuration. NASA TND-5164 (1969)
2. Chapman, A.J., Korts, H.H.: Free Jet Boundary with consideration of initial boundary layer. In: Proceedings of the Second US National Congress at Applied Mechanics, Univ. of Michigan, Ann. Arbur (1954)
3. Chuang, C.C., Chieng, C.C.: Supersonic base flow computation using higher-order closure turbulence models. J. Spacecrafts and Rockets 33(3), 374–380 (1996)
4. Friedrich, R., Ghosh, S., Mahle, I.: LES of compressible inert and reacting turbulent shear flows. In: Palma, I.M.L.M., Silva, A. (eds.) Invited lecture, Advances in Turbulence XI, Springer Proceedings in Physics, vol. 117, pp. 373–382. Springer, Heidelberg (2007)
5. Henckels, A., Gruhn, P.: Experimental Studies of Viscous Interaction Effects in Hypersonic Inlets and Nozzle Flow Fields. In: Jacob, D., Sachs, G., Wagner, S. (eds.) Basic Research Technologies for Two-Stage-to-Orbit Vehicles, pp. 383–403. Wiley-VCH, Chichester (2005)
6. Houtman, E.M., van der Weide, E.: Computational analysis of base flow/jet plume interaction. In: Proc. of the 3rd European Symp. on Aerothermodynamics for Space Vehicles, ESA SP-426 (1998)
7. Leer, V.: Towards the ultimate conservative difference scheme V. A second-order sequel to Godunov's method. J. of Computer Physics 32, 101–136 (1979)
8. Liou, M., Steffen, J.: A new flux splitting scheme. J. of Computational Physics 107, 23–39 (1993)
9. Meinke, M., et al.: A comparison of second- and sixth-order methods for large-eddy simulations. 31, 695–718 (2002)
10. Reace, A.J.: Turbulent flow predictions for after body nozzle geometries including base effects. J. Propulsion 7(3), 396–403 (1991)
11. Reig, J.L., Velázquez, A.: Quasi-Analytical Prediction of Base Flow-Plume Interaction. In: Proc. of the 3rd European Symp. on Aerothermodynamics for Space Vehicles, Noordwijk, ESA SP-426, pp. 599–604 (1998)
12. Reijasse, P., Delery, J.: Investigation of the Flow past the ARIANE 5 Launcher After body. J. of Spacecraft and Rockets 31(2), 208–214 (1994)
13. Sahu, J.: Numerical computations of supersonic base flow with special emphasis on turbulence modelling. AIAA J. 32(7), 1547–1549 (1994)
14. Smagorinsky, J.: General Circulation Experiments with the Primitive Equations. I. The Basic Experiment, Monthly Weather Review 91, 99–164 (1963)

15. Spalart, P., Allmaras, S.: A one-equation turbulence model for aerodynamic flows. Technical Report AIAA-92-0439, American Institute of Aeronautics and Astronautics (1992)
16. Wager, B., White, R.: Supersonic base flow problem in the presence of an exhaust jet. AIAA J. 18(8), 876–882 (1980)
17. White, F.M.: Viscous Fluid Flow. McGraw-Hill Book Company, New York (1974)
18. Wilcox, D.C.: Turbulence Modelling for CFD. DCW Industries, Inc., La Canada, California (1998)
19. Williamson, J.H.: Low-Storage Runge-Kutta Schemes. J. of Computational Physics 35, 48–56 (1980)

Application of Transpiration Cooling for Hot Structures

M. Kuhn and H. Hald

Deutsches Zentrum für Luft- und Raumfahrt e.V.
Institut für Bauweisen- und Konstruktionsforschung
Pfaffenwaldring 38-40, D-70569 Stuttgart, Germany
markus.kuhn@dlr.de

Summary

Specific parts of re-entry vehicles are exposed to severe conditions. Thereby, the material's capabilities can be exceeded by far and advanced cooling methods become necessary. Within the scope of this work, transpiration cooling was investigated in arc jet heated plasma flows by means of flat plate models. Screening tests pointed out, that transpiration cooling at the conditions tested is working well. Extensive testing at more severe conditions was done using three porous sample materials: Standard C/C with coolant flows parallel and perpendicular to the material's fibre layers and highly porous C/C. Coolant gases used were air, argon, helium and nitrogen. Minimal optimal coolant mass flows of 0.5 g/s Ar, 0.2 g/s He and 0.4 g/s N_2 were determined resulting in sample under surface temperature reductions of 50-60%. Altogether, sample under surface temperature reductions of 64% for He, 65% for Ar, 67% for air and 70% for N_2 were detected. These test series verified that transpiration cooling can be applied successfully for hot structures at application relevant re-entry conditions.

1 Introduction

1.1 Background

For highly demanded parts of re-entry vehicles, high temperature usable materials like ceramic matrix composites (CMC's) are being used. Such parts are in particular thermal protection system elements and hot structures like leading edges, flaps or components of propulsion systems. In most cases, the cooling of such hot structures relies on radiation cooling. However, in some situations the material's capabilities can be exceeded, for instance by higher-energetic (interplanetary) re-entry conditions, smaller structure nose radii or higher ballistic coefficients. Therefore, e.g. ablative or advanced cooling techniques like active cooling systems become necessary. Thereby, reusability is a great advantage of active cooling systems in contrast to ablative ones.

1.2 Motivation

Active cooling systems usually consist of internally, convective cooled structures like heat pipes. Additionally, it is possible to perform external active cooling systems; such systems could be for example transpired surfaces, which have not been applied successfully to reusable spacecraft yet. Basically, the principle of transpiration

A. Gülhan (Ed.): RESPACE – Key Tech. for Reusable Space Sys., NNFM 98, pp. 82–103, 2008.
springerlink.com

cooling consists of two heat mechanisms (see Fig. 1): Firstly, the porous structure is being cooled by convection of the coolant flow penetrating the porous media. Secondly, a thermal blocking coolant layer is built on the outer, hot surface of the porous structure, which tremendously reduces heat transfer to the outer surface.

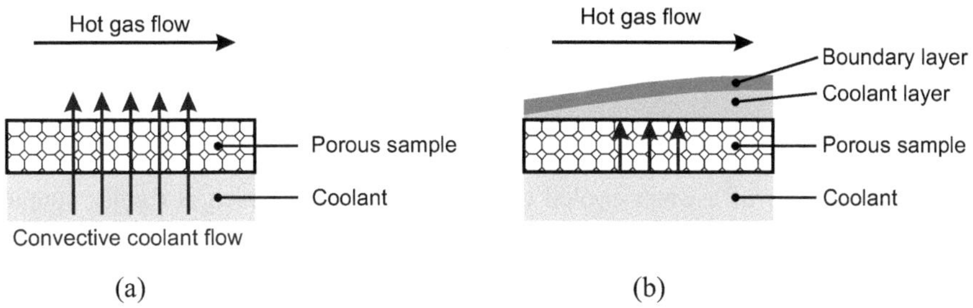

Fig. 1. Transpiration cooling mechanisms: (a) convective, (b) coolant layer

Already during the 50 s to 70 s, early studies of film, effusion and transpiration cooling have been done. Thereby, the main focus was laid on the cooling of gas turbine blades, rocket combustion chambers and hot structures like e.g. nose tip regions of intercontinental ballistic missiles. There are a lot of analytical and experimental approaches out of this time period, e.g. [9, 12, 13, 15, 16], but none of them contained experimental tests at low pressure supersonic plasma flow conditions. Although first considerations of such cooling methods for spacecraft existed during this time, no further investigations were made due to the lack of appropriate porous materials. In the 90's, those cooling techniques became interesting again and some detailed simulations were done, e.g. [1, 3, 8, 14, 19]. Since a couple of years, DLR is successfully working on transpiration-cooled rocket engines [2, 11, 18], which are fabricated out of porous CMC's. These materials are qualified candidates for transpiration cooling as they can be produced within a huge variety of open porosity and hence different permeability characteristics. Additionally, they exhibit excellent mechanical and thermal properties. Contrary to metal foams used until the 70s, CMC's do not fail if local hot spots occur. Metal foam structures tend to melt in the presence of local hot spots, which results in a whole structure failure. For more detailed information of DLR manufactured carbon- and oxide-based ceramics please refer to [6, 7, 17].

1.3 Proposed Aims

As there is a lack of experimental data for transpiration cooled CMC's in hypersonic plasma flows, the key aspects are laid on the experimental verification of such a cooling technique. The aim of this investigation is to estimate the effect of transpired surfaces for different gaseous coolant types and coolant mass flows as well as different sample materials. Those samples should be exposed to re-entry conditions in arc jet heated, hypersonic flows.

Therefore, already existing model holders should be modified and used for screening test series in facilities L2K and L3K [4, 5] in order to estimate the cooling influence of different coolants for different sample materials. Based on the experiences made there, a more detailed, final test series in the L3K facility should be performed. More detailed information of the testing facilities can be obtained in Chapter 7.

2 Design of Experimental Setup

2.1 Screening Test Series in L2K / L3K

Figures 2a and b show a basic setup of the models used in L2K and L3K. It is possible to incline the models at different angles of attack α in order to achieve variable heat loads. They consist of a water-cooled copper nose section, which is rigidly coupled with a water-cooled ground plate. Directly onto this plate, the coolant supply assembly is mounted, where different fed in coolant gases pass through a circular porous sample. Before and after the supply assembly, KAPYROK insulation material was placed, which consists of 91% aluminium oxide Al_2O_3 and 9% silicon oxide SiO_2. Thereby, heat transfer between the supply assembly and surrounding parts, particularly the water-cooled nose section should be avoided.

Figure 3 shows a schematic of the coolant supply assembly, forming the central part of the model. Basically, for both models used, the components are almost identical, though there are minor differences at some parts. A two-piece frame structure (2) made of stainless steel for L2K tests and of C/C-SiC for L3K tests, respectively, builds the skeletal structure for the assembly. Inside, a C/C-SiC cover plate (3) is attached, which was superimposed with a 100 μm thick, multilayered CVD-SiC coating for the L2K model to improve oxidation resistance. For the L3K model, no coating was applied. This cover plate holds a circular porous CMC sample (4), which is bonded with its conical section onto the conical counterpart of a reservoir tube made of C/C-SiC (5). The hereby used bond (*Polytec* 904) is based on ZrO_2 capable of resisting temperatures up to 2200°C. The reservoir tube is glued on an aluminium ring by means of an epoxy resin, which enables O-ring sealing to the supply rig (6) made of stainless steel. Using this rig, the coolant gas supply as well as two thermo-couples and a pressure sensor are being fixed. To support the bond between sample and reservoir tube, a spring mechanism is integrated to impose

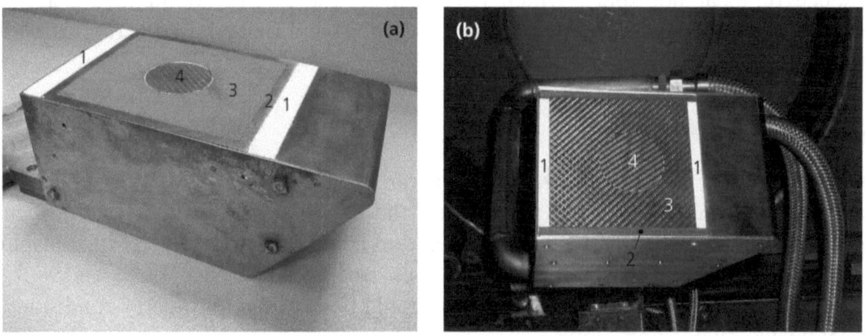

Fig. 2. Screening test models: (a) L2K model, (b) L3K model; 1 = KAPYROK insulation, 2 = frame structure, 3 = cover plate, 4 = porous sample

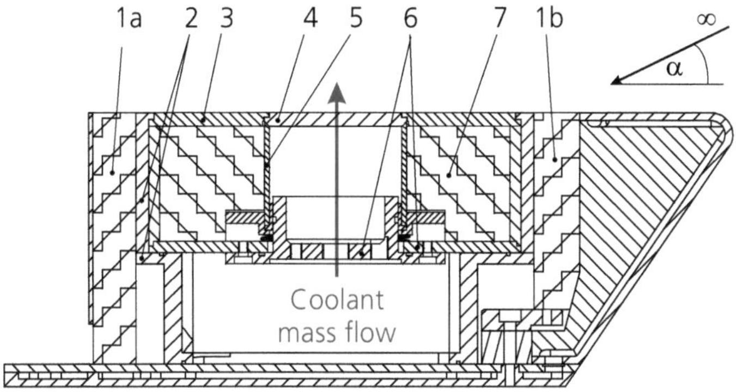

Fig. 3. Schematic of supply assembly for screening tests; 1a+b = KAYPROK insulation, 2 = frame structure, 3 = cover plate, 4 = porous sample, 5 = reservoir tube, 6 = supply rig, 7 = inner KAPYROK insulation

pressure upon the reservoir tube (5). Additionally, inside the setup KAPYROK material (7) is inserted in order to insulate the coolant reservoir against the surrounding parts.

At the beginning, first screening tests were carried out at L2K to check the model setup and to estimate appropriate cooling mass flow rates. Based on these experiences, a similar setup was made for L3K facility, which enables higher heat loads to the transpired samples. Due to different model holders in L2K and L3K, the geometric dimensions of the entire L2K model were 295 x 156 x 122 mm, whereas the L3K model was 73 mm shorter in length. Additionally, the centre of the CMC sample was 182 mm downstream of the stagnation point for L2K tests and 135 mm for L3K tests, respectively.

As can be seen in Fig. 4, inside the setup two thermocouples have been installed: The first one was bonded into a very small gap at the under side of the sample and measures the under surface temperature of the sample downstream. The other one measures the coolant gas temperature in the reservoir. Additionally, a pressure sensor made by *Kulite* records coolant reservoir pressures up to 3.4 bars. From outside, pyrometry has been used upstream of the sample, onto the sample (only L3K) at the position where thermocouple 1 is and shortly after the sample downstream (only L3K). Also, IR-thermography has been applied to observe the cooling behaviour, mainly directly on the sample and in the wake of the transpired region.

For the screening test series, four CMC sample materials with different porosities have been used, which can be seen in Table 1. All carbon based CMC's were manufactured by the Institute of Structure and Design, DLR Stuttgart and had an exit diameter of 60 mm and thickness of 6 mm; aluminium oxide based WHIPOX, manufactured by Institute of Materials Research, DLR Cologne, was 3 mm thick. Due to the smaller material thickness of WHIPOX, an appropriate C/C-SiC retainer ring was fabricated, into which the WHIPOX samples were bonded. This reduced the sample exit diameter to 44 mm for WHIPOX samples. Open porosities varied from 13 to 41%. It is remarked, that for C/C-SiC, the coolant flow occurs parallel to the fibre layers, while for the other samples the coolant flow passes perpendicular through to the fibre layers via micro cracks (see Fig. 5).

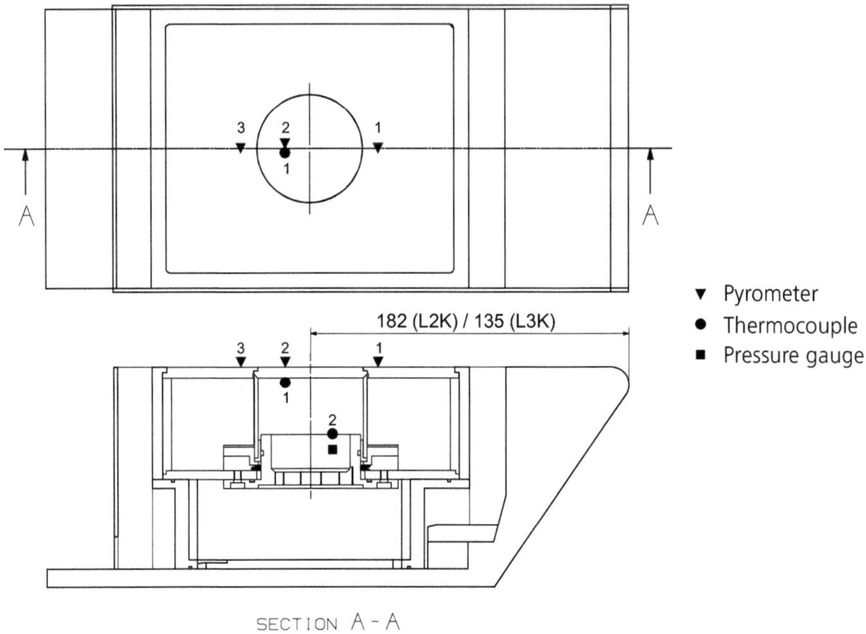

Fig. 4. Position of sensors for screening tests

Table 1. Sample materials for screening tests

Material	Coolant flow	*e'* [%]	*d* [mm]	Ø [mm]	*A* [m²]	Used in L2K	Used in L3K
C/C highly porous	$\perp$	41	6	60	$2{,}827\cdot10^{-3}$	×	×
C/C standard	$\perp$	13	6	60	$2{,}827\cdot10^{-3}$	×	×
C/C-SiC	$\parallel$	15	6	60	$2{,}827\cdot10^{-3}$	×	
WHIPOX	$\perp$	33	3	44	$1{,}521\cdot10^{-3}$	×	×

e' open porosity, *d* thickness, Ø exit diameter, *A* exit area.

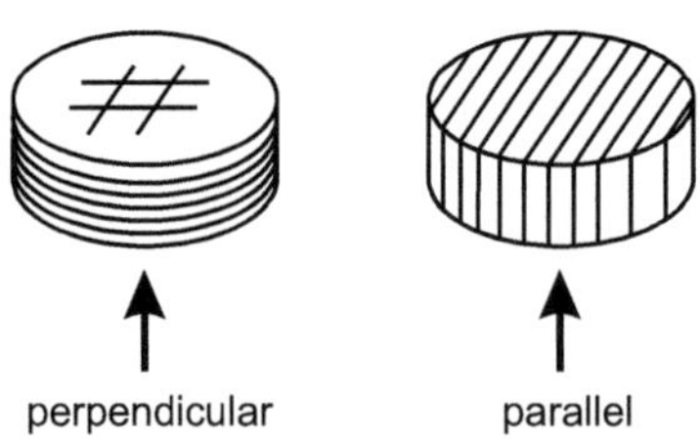

Fig. 5. Coolant flow direction relating to sample fibre layers

2.2 Main Test Series in L3K

Many experiences have been made at the screening test series; especially sealing problems have been detected as major challenges. At high heat loads for instance, different thermal expansion characteristics of the reservoir tube and the glued on aluminium ring caused the reservoir tube to break. Supplementary, rapid increases of heat fluxes onto the sample – and hence transferred heat to the bonded joint between reservoir tube and sample – caused the ceramic bond to break due to insufficient resistance to thermal shocks. From this follows that a leakage of the coolant reservoir makes further testing useless.

Therefore, the setup has been modified at the appropriate parts. Figures 6a and b show the model setup for the main test series conducted in L3K. Just like at the previous models, it is possible to incline the model at different angles of attack in order to achieve variable heat loads. Regarding the setup, the model bases upon a water-cooled copper nose section rigidly coupled with a water-cooled ground plate. Onto this plate, the coolant supply assembly is mounted, where different fed in coolant gases pass through a rectangular porous sample embedded in a cover frame. Around the supply assembly, KAPYROK insulation material was inserted in order to avoid heat transfer between the outer, hot parts and the inner, water-cooled parts.

Figure 7 shows an exploded view of the coolant supply assembly for the main test model. The whole assembly is fixed to a substructure made of copper and stainless steel (8), which is installed onto the ground plate of the model holder itself. Directly onto this substructure, the coolant reservoir (7) is being placed, whereas at the reservoir sidewalls the coolant supply and sensors are attached. On top of the reservoir, the porous sample (6) and a graphite seal (5) are placed. In the following, a metallic sealing unit (4) made of *Plansee* PM 2000 material is attached – this part can be screwed together with the substructure which effects, that the porous sample can be sealed against the coolant reservoir. To avoid heating of the metallic sealing unit, insulation (3) made of aluminium oxide fleece was positioned between the sealing unit (4) and cover frame (2) made of C/C-SiC. Around the cover frame and porous probe a C/C-SiC cover plate (1) is attached, which was superimposed with a 100 μm thick, multilayered CVD-SiC coating to improve oxidation resistance.

For this model, another holder has been used which effected in smaller height and greater width; geometric dimensions of the model were 285.5 x 194 x 60 mm. Thereby, the centre of the samples was positioned 107.1 mm downstream of the stagnation point and 40 mm distant to the model centreline. Contrary to previous models, the samples used here were rectangular with exit dimensions of 61 x 61 mm and they were placed at

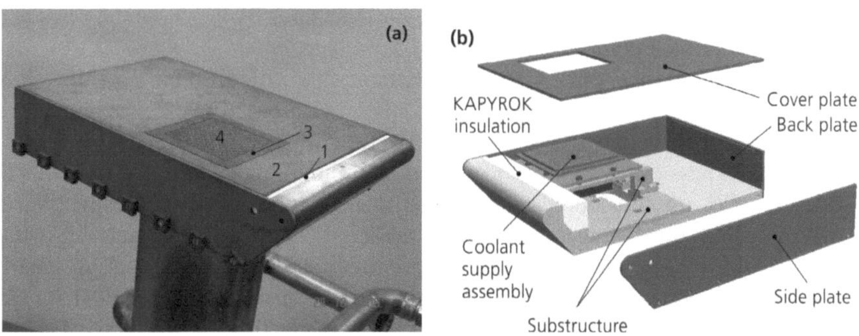

Fig. 6. (a) Main test model and its (b) exploded view (without inner insulation), 1 = KAPYROK insulation, 2 = cover plate, 3 = cover frame, 4 = porous sample

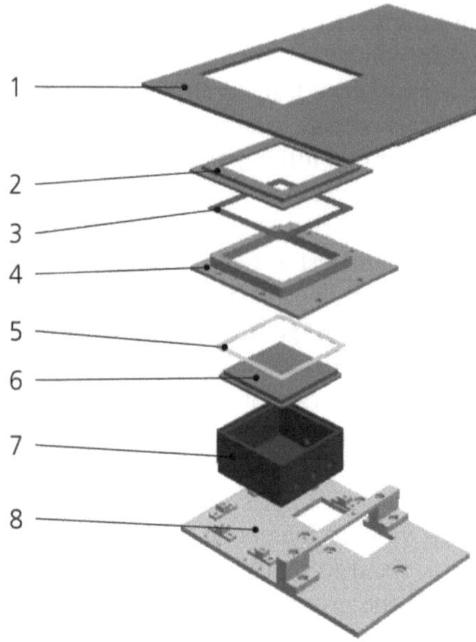

Fig. 7. Schematic of supply assembly for main tests; 1 = cover plate, 2 = cover frame, 3 = insulation, 4 = sealing unit, 5 = graphite seal, 6 = porous sample, 7 = coolant reservoir, 8 = substructure

one model half only, as can be seen in Fig. 8. This results in a cooled and uncooled side, which makes direct estimations of the cooling effects possible.

Furthermore, additional sensors have been installed as can be seen in Fig. 8. Besides the pressure measurement in the coolant reservoir, it is possible to measure coolant gas temperature and sample under side temperature by means of a thermocouple bonded into a small gap in the sample. At the same position of the sample on top, a pyrometer spot was placed, which records the sample surface temperature. To compare the cooling effect, two pyrometers were placed on the uncooled side (with same distance to the stagnation point as the sample pyrometer spot) and two more pyrometers upstream of the sample. Twelve additional thermocouples were attached in different heights inside the setup in order to estimate the cooling effect onto the internal structure. Furthermore, IR-thermography has been applied to observe the cooling behaviour.

For the main test series, three CMC sample materials with different porosities have been tested, which can be seen in Table 2. Only carbon based CMC's manufactured by the Institute of Structure and Design, DLR Stuttgart with exit dimensions of 61 x 61 mm and thickness of 6 mm have been used; aluminium oxide based WHIPOX [17] and OXIPOL [7] are considered to be tested in near future. Open porosities varied from 16 to 44%. Extensive testing has been made with standard C/C material, which can be produced very well regarding reproducibility and quality. Additionally, stratified C/C based on standard C/C material has been used. It is marked by C/C standard ∥, because coolant flow occurs parallel to the fibre layers here. The last material tested was highly porous C/C.

Table 2. Sample materials for main tests

Sample material	Coolant flow	*e'* [%]	*d* [mm]	*x·y* [mm]	*A* [m²]
C/C standard ⊥	⊥	17-18	6	61·61	$3{,}721 \cdot 10^{-3}$
C/C standard ‖	‖	16	6	61·61	$3{,}721 \cdot 10^{-3}$
C/C highly porous	⊥	44	6	61·61	$3{,}721 \cdot 10^{-3}$

e' open porosity, *d* thickness, *x·y* exit dimensions, *A* exit area.

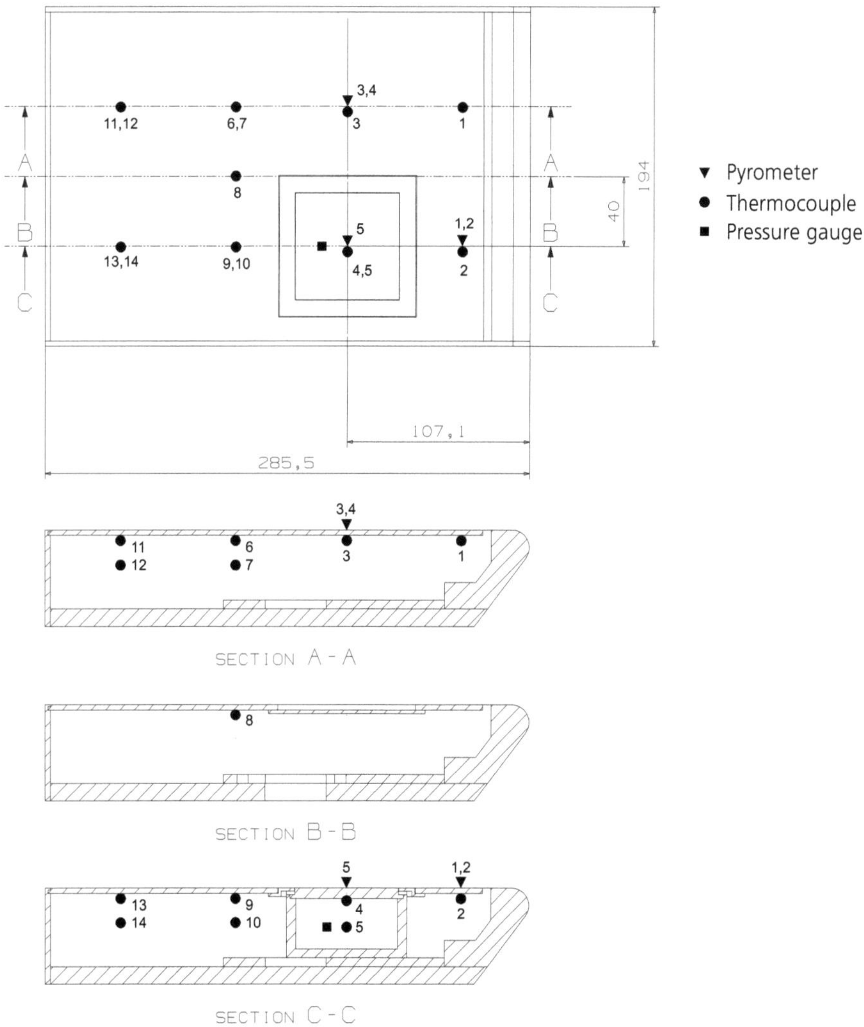

Fig. 8. Position of sensors for main tests

3 Experimental Results

A detailed description of the experimental facilities can be seen in Chapter 7.

3.1 Screening Tests

At the beginning, first tests were conducted at relatively moderate enthalpies at L2K facility. Afterwards, a second test series was carried out at L3K facility at higher enthalpies and hence higher surface temperatures. The more interested reader is referred to [10]. Flow conditions for both test series are shown in Table 3.

It was considered to compare different coolant gases based on equal volume flows. At operation, the coolant mass flows were adjusted proportional to the coolant gases densities at ambient conditions. Hence, e.g. a mass flow rate of 1.00 g/s N_2 would equal 0.14 g/s He or 1.43 g/s Ar.

Table 3. Flow conditions of screening tests

	L2K	L3K FC-II
Reservoir pressure [hPa]	1000	4550
Reservoir temperature [K]	4220	5400
Total enthalpy [MJ/kg]	8.4	11.0
Mass flow [g/s]	36	142
Nozzle exit diameter [mm]	200	300
Model position downstream [mm]	250	300
Free stream Mach number [-]	7.0	7.6
Free stream static pressure [Pa]	47	50
Free stream static temperature [K]	378	491
Angle of attack [°]	20	30
Free stream velocity [m/s]	2996	3730
Mass fraction N_2	0.739	0.757
Mass fraction O_2	0.046	0.012
Mass fraction NO	0.045	0.018
Mass fraction N	$< 10^{-6}$	$< 10^{-6}$
Mass fraction O	0.170	0.213

Screening tests L2K

Main goals of this attempt were to determine, which sample materials are suited for this kind of cooling. Additionally, different coolant gases should be used in order to determine efficiency of the gas itself and the required mass flows. All tests were performed by inserting the model without cooling gases switched on and thereby heating up the whole configuration. After reaching sample under surface temperatures of about 700 K, the coolant gases where switched on and increased stepwise. Coolant gases used for these tests were N_2, He, Ar and air at typical test times of about 400 s. Maximum coolant gas pressures recorded in the reservoir were about 3.0 bars for standard C/C, 1.1 bars for highly porous C/C, 1.6 bars for WHIPOX and 0.2 bars for C/C-SiC.

Figure 9 shows typical temperature traces of the cover plate (upstream of the sample), coolant gas and the sample under surface for a WHIPOX sample and different nitrogen coolant mass flows. Due to short test times, no steady state conditions could be reached. Nevertheless, the temperature at the lower sample surface was assumed to be quasi-stationary at the end of each mass flow step. A further unfavourable effect is the rising of coolant gas temperature by time; this can be explained by the heating of the whole setup from top, which heats up the coolant gas as well. Hence, the reduction of sample under surface temperatures will not be that effective after several small coolant mass flow steps as if the final coolant mass flow would be switched on directly.

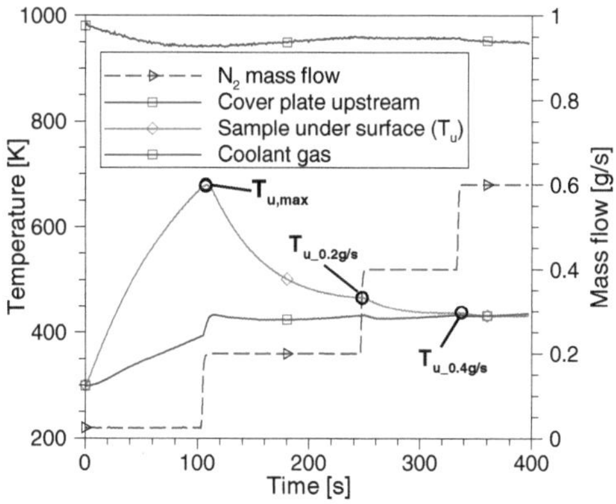

Fig. 9. Typical temperature traces at $\alpha = 20°$, coolant: N_2, sample: WHIPOX

To estimate the cooling efficiency, these quasi-stationary values were used. Usually, for film cooling comparability, the cooling efficiency η is defined by the ratio of Stanton numbers or heat fluxes for the cooled (index c) in relation to the uncooled case (index 0)

$$\eta(F) = 1 - \frac{St_c}{St_0} = 1 - \frac{\dot{q}_c}{\dot{q}_0} \tag{1}$$

This efficiency is dependent on the blowing ratio F, which is defined by the ratio of ρu for the coolant (index c) in relation to the flow conditions at the outer edge of the boundary layer (index e)

$$F = \frac{(\rho u)_c}{(\rho u)_e} \tag{2}$$

Determination of heat fluxes according to Eq. 1 for this kind of plasma flows is difficult to perform due to high temperature effects. This problem is well known and will have to be investigated more in future. Furthermore, the integration of a heat flux sensor into the porous sample would be unfavourable as it would disturb the coolant flow through the porous sample.

Additionally, a detached, curved shock wave forms at the nose part of the holder. Unfortunately, the flow conditions after the shock at the sample position is not known. Also, the effective open pore area of the porous samples was not known, from which the coolant velocity u could be calculated by means of continuity. Hence, the blowing ratio F (Eq. 2) is not formed here and the coolant mass flow is used therefore. Instead of η, an alternative comparability number Z is defined, which describes the temperature decrease at the sample under surface in comparison to the maximum occurring temperature

$$Z = \frac{T_{u,\max} - T_{u,\dot{m}}}{T_{u,\max}} \cdot 100\% . \tag{3}$$

Hereby, $T_{u,\dot{m}}$ are the temperatures at the sample under surface after each coolant mass flow step (see Fig. 9). $T_{u,max}$ is the maximum occurring temperature at the end of the heating phase and acts as reference temperature, which is between 450 and 760 K.

Figures 10a-d show the comparability numbers Z for all samples and coolant gases tested. It can be seen, that there is a different trend for each coolant gas at same gas

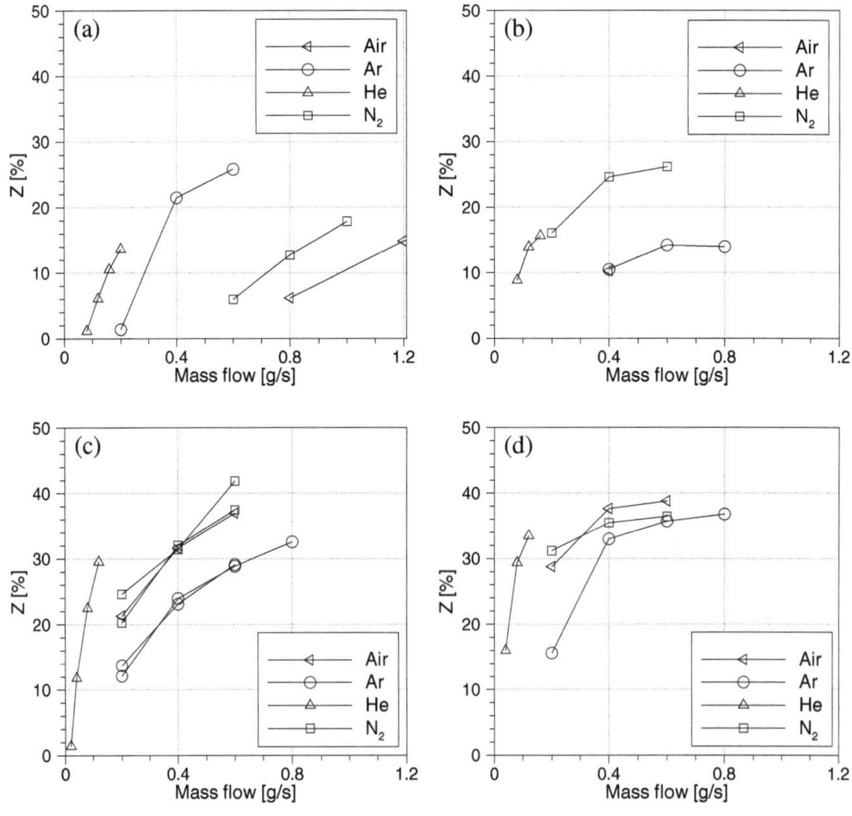

Fig. 10. Comparability number Z at $\alpha = 20°$ for (a) C/C standard, (b) C/C highly porous, (c) C/C-SiC, (d) WHIPOX

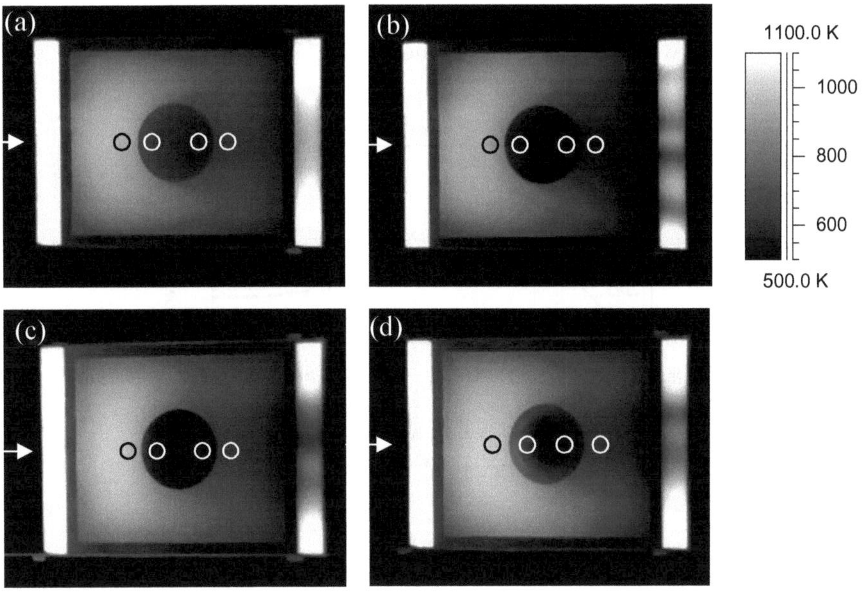

Fig. 11. IR-thermography for nitrogen mass flows of 0.4 g/s at $\alpha = 20°$; (a) C/C standard, (b) C/C highly porous, (c) C/C-SiC, (d) WHIPOX

volume flows. N_2 seems to be the most efficient coolant of all followed by air and Ar. Contrary to all other coolants, He turned out to be less effective. This leads to the assumption that coolant properties like heat conductivity or capacity have to be considered more detailed in future. There is only one exception that exists for standard C/C (Fig. 10a). Ar seems to be more effective as N_2 or air.

Taking a closer look on the different porous sample materials, it can be concluded, that for highly permeable samples like C/C-SiC and WHIPOX, the efficiency is increased significantly. But is has to be remarked, that for C/C-SiC, the fibre layers are in coolant flow direction and WHIPOX had a thickness of only 3 mm. One explanation could be the holding time of the coolants while passing the porous sample. This time will be less as for highly porous or standard C/C, hence the convective heat transfer in the structure will be lower, but the thermal blocking effect in the boundary layer higher as the coolant enters the boundary layer slightly cooler.

In total, maximum temperature reductions Z of 34% for helium, 36% for argon, 38% for air and 42% for nitrogen are achieved.

For the following investigations, all different sample materials are compared with N_2 cooling. In order to see, how homogeneous the coolant blowing is, IR-thermography has been applied, which can be seen in Fig. 11 for nitrogen mass flows of 0.4 g/s.

Hereby, the flow direction was from the left. It can clearly be seen, that the cooling gas significantly reduces the sample temperatures in contrast to the cover plate upstream. It seems that the temperature decrease is greatest for highly porous C/C. However, temperature irregularities can be seen on top of the sample. This indicates, that the coolant gas escapes inhomogeneous, which can also be seen in three darker stripes at the KAPYROK insulation material downstream. These KAPYROK bars are white in colour due to different surface emissivity characteristics.

For standard C/C, blowing is very uniform; sample surfaces seem to be nearly isothermal and the wake region at the position of the KAPYROK blocks looks very homogeneous showing no darker stripes.

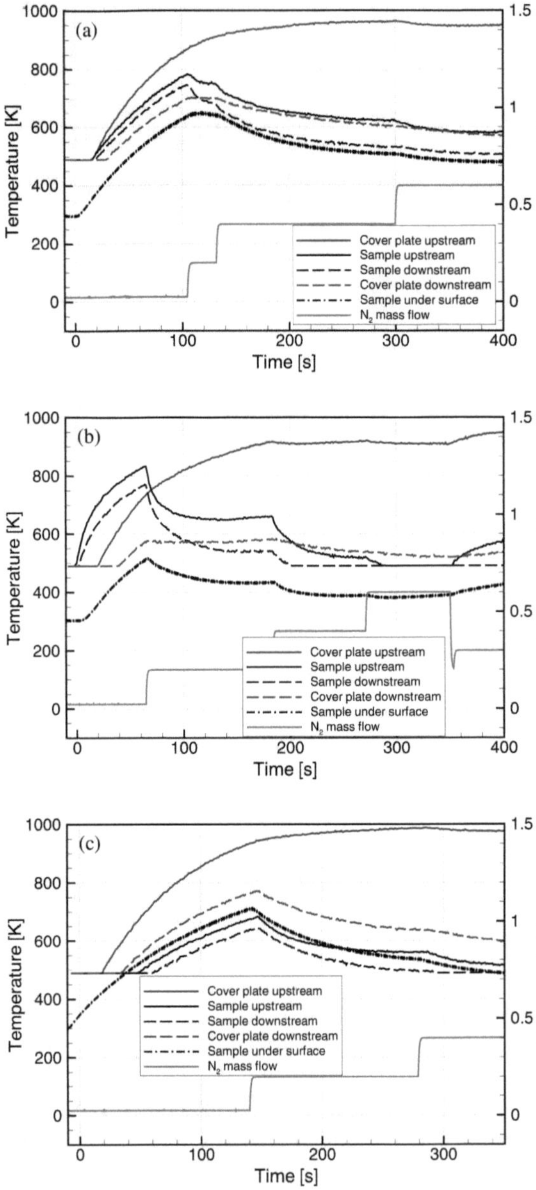

Fig. 12a-c. IR-thermography spots for coolant N_2 at $\alpha = 20°$; (a) C/C standard, (b) C/C highly porous, (c) C/C-SiC

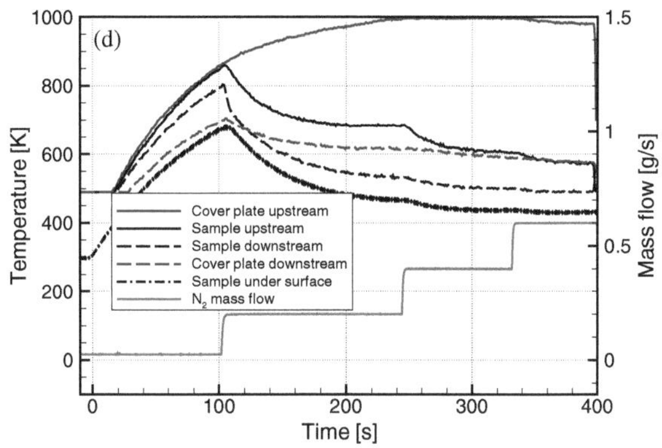

Fig. 12d. IR-thermography spots for coolant N_2 at $\alpha = 20°$; (d) WHIPOX

C/C-SiC and WHIPOX appear to have a more homogeneous blowing at the sample surfaces, though there are minor inhomogeneities visible.

It is to remark, that no significant flow field disturbances can be seen herby.

Figures 12a-d show some temperature line plots for selected spots on the middle axis for each material and N_2 coolant mass flows. These spots accord to the indicated circled spots in Figs. 11a-d. It can be seen, that the temperatures decrease over the length of the samples. Also to be noticed is the temperature decrease of the embedding C/C-SiC cover plate up- and downstream – whereby the cooling effect of the transpired samples even influences the temperatures of the cover plate upstream.

For C/C-SiC material, the sample under surface temperature seems to be higher than for the other sample materials. Possibly, due to high permeability of this material, hot gas is entering into the reservoir through the porous sample during the heating phase and so heating up the sample. Secondary, slightly uncertainties in sample emissivity characteristics and absorptance of the optical windows could explain this effect.

Screening tests L3K

The second screening test series in L3K should verify the experiences made in L2K at application relevant conditions and hence higher surface temperatures. This was mainly achieved by increasing the total enthalpy of the free stream, the static pressure at the model position and the angle of attack. Table 3 shows the flow condition "L3K FC-II" used hereby. Coolant gases used for L3K experiments were N_2 and He at typical test times of 180-220 s.

Again, the model was inserted without cooling gases switched on and thereby heating up the whole configuration. After reaching sample under surface temperatures of about 800-1000 K, the coolant gases where switched on in one step to coolant mass flows of 1.0 g/s for N_2 and 0.2 g/s for He. Sample materials tested were standard C/C, highly porous C/C as well as WHIPOX. Just like at L2K screening tests, N_2 turned out to be more effective than He [10].

Due to the higher surface temperatures, the internal supply assembly was heated up faster. This caused the coolant reservoir and hence coolant gas temperature heating up very quickly, which resulted in shorter test times. However, significant sealing problems at the coolant gas reservoir occurred.

Higher heating loads effected that the reservoir liner broke and a leakage developed, as can be seen in Fig. 13. Due to different thermal expansion characteristics of the reservoir tube and the glued on aluminium ring a crack in longitudinal direction was observed. The brownish colouration inside the C/C-SiC liner arises from silicone sealing material, which was tried to apply in order to seal the material.

The second weak point of this setup was the bonded joint between reservoir tube and sample. The rapid surface temperature increase after insertion of the model caused the ceramic bond to break and build micro cracks due to insufficient resistance to thermal shocks.

The problems mentioned yielded in a stop of this test series. Though the measurement results obtained were not quite good, the know-how regarding sealing problems gained here was worth a lot for prospective model setups.

Fig. 13. Broken coolant reservoir at L3K screening tests

3.2 Main Tests L3K

Based on the experiences made at the two screening test series, a new model had been built. Critical parts like the coolant reservoir had been modified and fabricated out of C/C-SiC in two versions. One reservoir was made by means of castings, while the other one was fabricated out of thicker C/C-SiC plate material. Sealing of the reservoir was managed by a graphite seal. The modifications did work well, so that the more severe flow condition "L3K FC-III" was used at 20° and 30° angle of attack, which is detailed in Table 4. Coolant gases used for these tests were N_2, He, Ar and air at typical test times of about 180-300 s, which is characterised by greyish shaded boxes in the following diagrams. Also, several test runs have been performed to check repeatability, which was in very good agreement.

Again, it was considered to compare different coolant gases based on equal volume flows. In contrast to the screening tests, the model was inserted with cooling gases switched on; in addition, only a fixed coolant mass flow was adjusted.

Table 4. Flow condition of main tests

	L3K FC-III
Reservoir pressure [hPa]	4700
Reservoir temperature [K]	5650
Total enthalpy [MJ/kg]	11.6
Mass flow [g/s]	142
Nozzle exit diameter [mm]	300
Model position downstream [mm]	300
Free stream Mach number [-]	7.5
Free stream static pressure [Pa]	56
Free stream static temperature [K]	530
Angle of attack [°]	20 / 30
Free stream velocity [m/s]	3873
Mass fraction N_2	0.7620
Mass fraction O_2	0.0040
Mass fraction NO	0.0064
Mass fraction N	0.0001
Mass fraction O	0.2270

In comparison to the screening test samples, the exit area of the main test samples was increased by 32%. Based on the detected (screening tests) coolant mass flows required for satisfying cooling effects, the coolant mass flows were also increased by approximately 30% at the beginning. Thereby, it turned out that the increased mass flows were too high. This indicates that the sealing mechanism of the new model setup works much better than at the screening tests; the entire coolant gas passes through the sample now, whereas at the screening tests minor leakages existed. To detect the leakage flow in the coolant reservoir, an impermeable metallic sample had been installed and the reservoir was pressurized. The leakage flows determined by pressure loss were 3‰ for N_2 and 4‰ for He, which can be neglected.

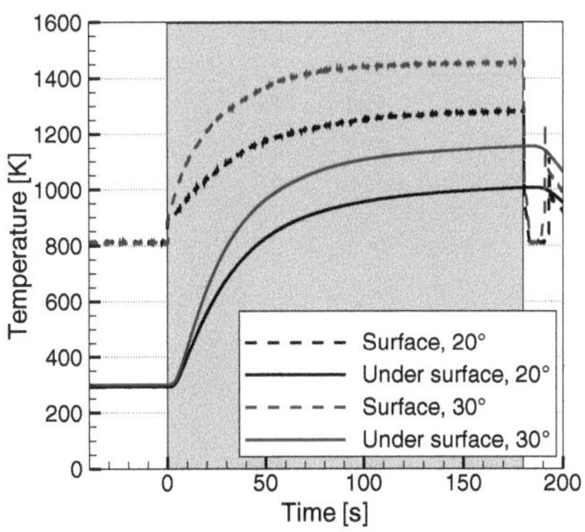

Fig. 14. Uncooled sample temperatures for C/C standard ⊥ at α = 20° / 30°

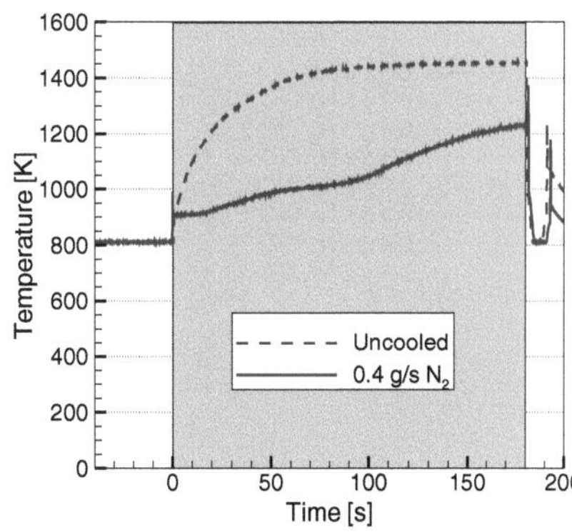

Fig. 15. Surface temperatures for standard C/C ⊥ at α = 30°; uncooled and for a nitrogen mass flow of 0.4 g/s

In the following, only details of the porous sample are discussed, which indicates the feasibility of transpiration cooled, hot structures. More detailed information of the other measurements can be seen in Chapter 7.

Although the model exhibits an uncooled half as reference, two tests without cooling have been performed at angles of attack α of 20° and 30°, which can be seen in Fig. 14. It shows the surface temperature measured by a pyrometer and the under surface temperature measured by a thermocouple for standard C/C ⊥ material. For $\alpha = 20°$, temperatures of about 1300 K for the surface and 1000 K for the under surface were measured; for $\alpha = 30°$, temperatures of about 1450 K for the surface and 1150 K for the under surface were measured.

With cooling, sample surface temperatures will be highly reduced, so that the temperatures are below the pyrometer's measurement range. Only for the uncooled cases at $\alpha = 20°$ and 30° and for a nitrogen mass flow of 0.4 g/s at $\alpha = 30°$, the surface temperatures where in the measurement range. Figure 15 shows the sample surface temperature measured by pyrometer for standard C/C ⊥ and the cases mentioned at $\alpha = 30°$. Here, it seems that a N_2 mass flow of 0.4 g/s is too low as the temperature doesn't reach stationary conditions and rises farther.

In order to compare the main tests with the screening tests, only under surface temperatures measured by thermocouple are discussed next. For C/C standard ⊥ samples, a series labelled by C/C-1 to C/C-3 originating from the same manufactured plate material was used thereby.

As can be seen in Fig. 16, for $\alpha = 20°$ no stationary conditions were achieved, but the influence of different mass flow rates can clearly be seen. While the reduction of a N_2 mass flow of 0.3 g/s seems to be too low, the effect of higher mass flows appears to be getting smaller. For $\alpha = 30°$ (see Fig. 16b), the temperatures are just slightly higher, although a higher angle of attack effects higher heat loads. Here, it seems that a N_2 mass flow of 0.4 g/s is too low as the temperature rises unproportional in contrast to higher mass flows. The cooling effect of N_2 and air is in good agreement for both angles of attack as expected due to similar physical properties.

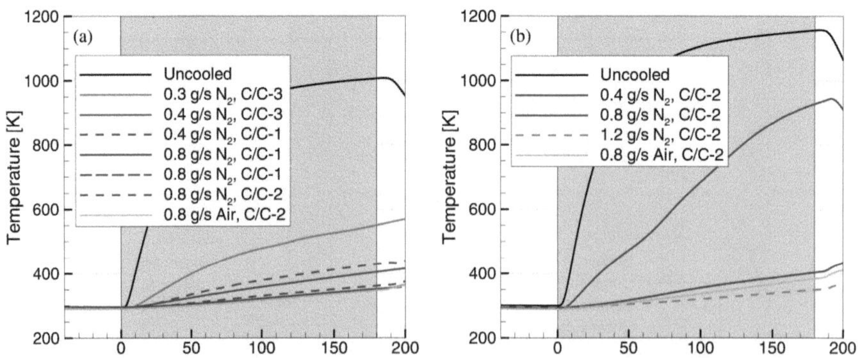

Fig. 16. Under surface temperatures for standard C/C ⊥ and coolants air and N_2, (a) $\alpha = 20°$ and (b) $\alpha = 30°$

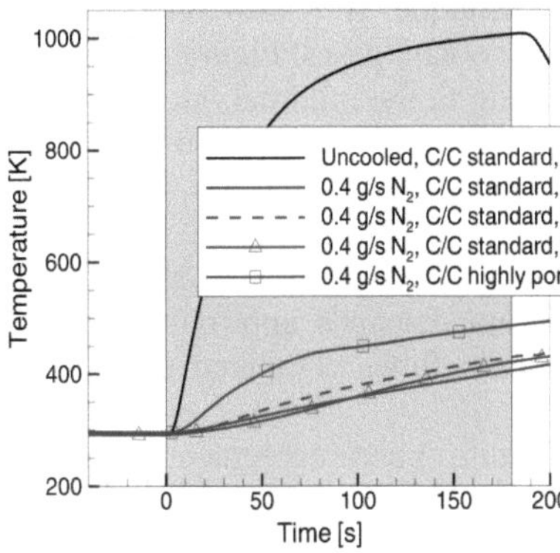

Fig. 17. Under surface temperatures for a nitrogen mass flow of 0.4 g/s and different sample materials at $\alpha = 20°$

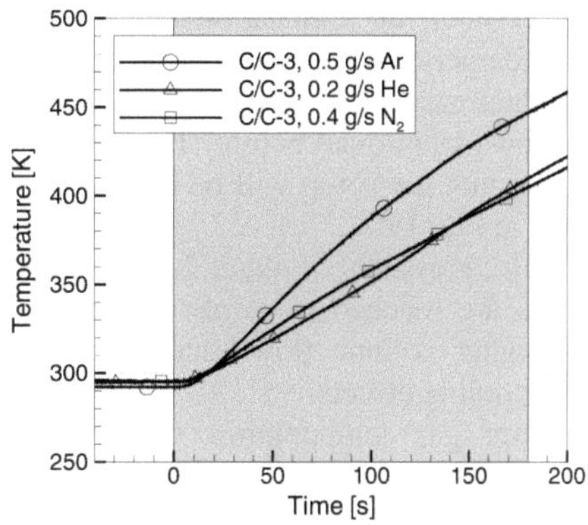

Fig. 18. Under surface temperatures of standard C/C ⊥ for ideal coolant mass flows at $\alpha = 20°$

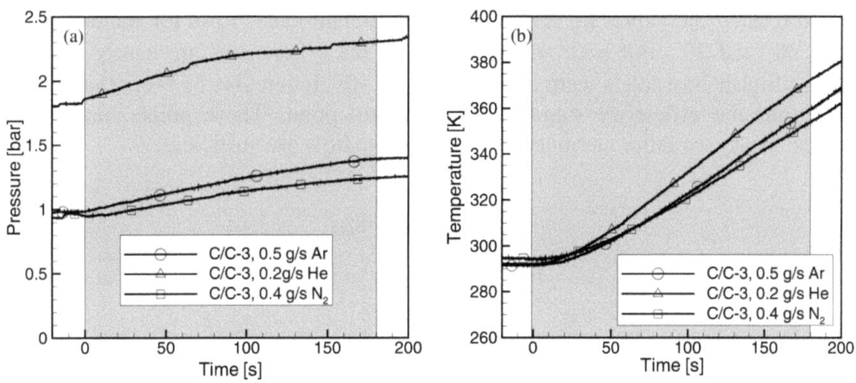

Fig. 19. (a) Reservoir pressures and (b) coolant gas temperatures for ideal coolant gases and standard C/C ⊥, $\alpha = 20°$

Based on the experiences made at $\alpha = 20°$, N_2 mass flows of 0.4 g/s turned out be effective enough at this angle of attack. Therefore, sample materials standard C/C ⊥ and ∥ as well as highly porous C/C have been tested at this mass flow rate, see Fig. 17. The temperature history for both standard C/C materials looks very similar; whereas for highly porous C/C the temperature increases more at the beginning, but turns later to rise with a similar gradient as the other materials.

Adapted from ideal nitrogen mass flows of 0.4 g/s at $\alpha = 20°$, several tests have been conducted in order to find out, which equivalent mass flows are demanded for coolant gases Ar and He. In doing so, Ar mass flows of 0.5 g/s and He mass flows of 0.2 g/s have been detected, which gain similar satisfying cooling effects. Figure 18 shows the under surface temperatures for standard C/C ⊥ and the different coolant mass flows mentioned.

Concerning the application of such a cooling technique, it is also important to know the operating pressures and temperatures of the coolant gases. Figure 19a shows the coolant gas pressures in the reservoir corresponding to the minimal, ideal coolant mass flows mentioned before. The corresponding coolant pressures tend to rise at the tests, because the setup will be heated up from top and so the coolant gas, which can be seen in Fig. 19b.

For He, pressures of about 2.3 bars, while for N_2 and Ar, pressures of about 1.2 – 1.4 bars are reached. It is also remarked, that helium demands approximately 3.5 times higher volume flows than nitrogen for the mass flows compared to achieve similar cooling effects.

Coolant gas temperatures of Ar and N_2 in quite good agreement, He gas temperatures tend to be slightly higher than the others. Ideally, the gas temperatures should be constant, which cannot be avoided at this kind of hot structures due to internal heating. On the other side, high coolant gas flows could eliminate this problem, but this is not desired as the weight for a real spacecraft will be increased and the boundary layer will be disturbed more.

Just as for the screening tests, a comparability number *Z* (which characterises the temperature reduction on the sample under surface) was determined by Eq. 1. Figures 20a and b show the values for *Z* dependent on coolant mass flows for standard C/C ⊥ at $\alpha = 20°$ and 30°. For both angles of attack, the efficiencies are nearly the same, although higher heat loads were applied for $\alpha = 30°$. It can also be seen, that for each coolant gas the efficiency stagnates at a certain point. These points out, that for technical aspects smaller amounts of coolant mass flow are sufficient.

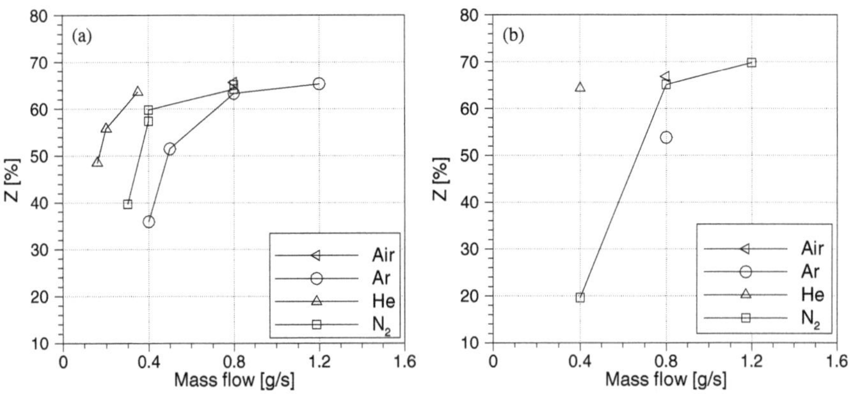

Fig. 20. Comparability number Z for C/C standard ⊥ and different coolant gases, (a) $\alpha = 20°$, (b) $\alpha = 30°$

He seems to be very sufficient compared by mass flows, but compared by volume flows, helium demands much higher volume flows than nitrogen to achieve equivalent cooling. N_2 and air are nearly equal, which was expected due to similar physical properties. Ar turned out to be less effective at same volume flows as already been observed at the screening tests.

Altogether compared at same mass flows, He is most efficient, followed by N_2 and air and Ar at the end. For Z, maximum values of 64% for helium, 65% for argon, 67% for air and 70% for nitrogen are achieved.

Figure 21 illustrates the comparability number Z for nitrogen mass flows of 0.4 g/s at an angle of attack of 20°. Contrary to the screening tests, the values for highly porous C/C are slightly lower than for standard C/C. Thereby, the Z values of C/C standard ‖ and C/C highly porous correspond to maximum temperatures measured for C/C standard ⊥ material.

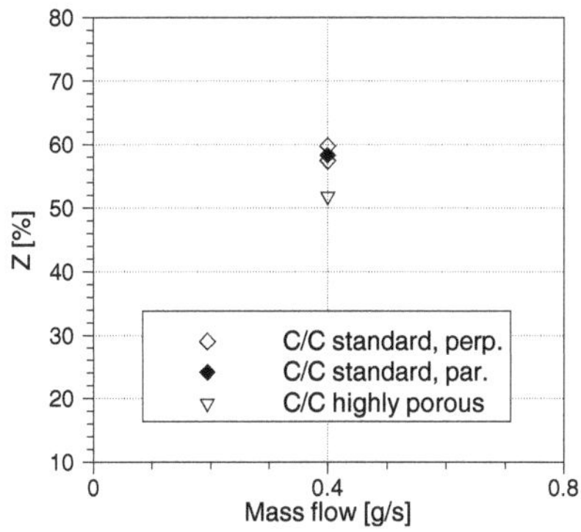

Fig. 21. Comparability number Z for nitrogen mass flows of 0.4 g/s and different sample materials at $\alpha = 20°$

4 Summary

Screening tests conducted in the L2K facility showed, that transpiration cooling at the conditions tested is working well. Sample materials used included C/C standard, C/C highly porous, C/C-SiC and WHIPOX and were tested for coolant gases air, argon, helium and nitrogen. For C/C-SiC and WHIPOX, a higher decrease in under surface temperature was observed, whereat C/C-SiC was based on stratified material and WHIPOX samples were only half as thick as all others. The most homogeneous blowing properties were observed for standard C/C. In total, maximum under surface temperature reductions, characterised by Z, of 34% for helium, 36% for argon, 38% for air and 42% for nitrogen were achieved.

Afterwards, a second screening test series in L3K was performed to verify the cooling setup at application relevant conditions. It turned out, that the setup used was not qualified for the more severe flow conditions here. Major sealing problems have been detected, which were very valuable for further investigations.

A new model had been designed and successfully tested in L3K at an even more severe flow condition and two angles of attack (20/30°). Leakage flows of about 3-4‰ were determined, which can be neglected. Three materials, standard C/C with coolant flows parallel and perpendicular to the material's fibre layers and highly porous C/C, were tested. Coolant gases used were air, argon, helium and nitrogen. Although the sample exit area was enlarged by 32%, the appropriate coolant mass flows detected to achieve satisfying cooling effects were much lower compared to the screening tests. Thereby, minimal ideal coolant mass flows of 0.5 g/s Ar, 0.2 g/s He and 0.4 g/s N_2 were determined; for this mass flows, Z values were between 50-60%. Altogether, under surface temperature reductions Z of 64% for helium, 65% for argon, 67% for air and 70% for nitrogen were achieved.

These test series verified that transpiration cooling can be applied successfully for hot structures at application relevant re-entry conditions. For all tests performed, no significant disturbances of the flow fields have been observed.

To compare the cooling efficiency more precisely, investigations of e.g. effective sample pore exit area, incoming heat fluxes onto the sample and free stream conditions at the sample position have to be done. This would enable the determination of cooling efficiency $\eta(F)$, which is commonly used in the literature.

References

1. Belletre, J., Bataille, F., Lallemand, A.: A New Approach for the Study of Turbulent Boundary Layers with Blowing. International Journal of Heat and Mass Transfer 42, 2905–2920 (1999)
2. Ghadiani, S.R.: A Multiphasic Continuum Mechanical Model for Design Investigations of an Effusion-Cooled Rocket Thrust Chamber. Ph.D. thesis, Universität Stuttgart, Germany (2005)
3. Glass, D.E., Dilley, A.D., Kelly, H.N.: Numerical Analysis of Convection / Transpiration Cooling. Journal of Spacecraft and Rockets 38, 15–20 (2001)
4. Gülhan, A., Esser, B.: Arc-Heated Facilities as a Tool to Study Aerothermodynamic Problems of Reentry Vehicles. In: Lu, F.K., Marren, D.E. (eds.) Advanced Hypersonic Test Facilities, Progress in Astronautics and Aeronautics, vol. 198, pp. 375–403. AIAA (2002)
5. Gülhan, A., Esser, B., Koch, U.: Experimental Investigation on Local Aerothermodynamic Problems of Re-entry Vehicles in the Arc Heated Facility LBK. Journal of Spacecrafts and Rockets 38, 199–206 (2001)
6. Heidenreich, B.: Herstellung von Faserkeramiken nach dem Flüssigsilicierverfahren (LSI-Technik). In: Krenkel, W. (ed.) Keramische Verbundwerkstoffe, pp. 48–75. Wiley-VHH, Weinheim (2003)
7. Heidenreich, B., et al.: Net Shape Manufacturing of Fabric Reinforced Oxide/Oxide Components via Resin Transfer Moulding and Pyrolysis. In: Proceedings of 28th International Cocoa Beach Conference and Exposition on Advanced Ceramics & Composites, Cocoa Beach, USA, January 25-30 (2004)
8. Hwang, C.B., Lin, C.A.: Low-Reynolds number k-epsilon Modelling of Flows with Transpiration. International Journal for Numerical Methods in Fluids 32, 495–514 (2000)
9. Kays, W.M.: Heat Transfer to the Transpired Turbulent Boundary Layer. International Journal of Heat and Mass Transfer 15, 1023–1044 (1972)
10. Kuhn, M., et al.: Experimental Investigations of Transpiration Cooled CMC's in Supersonic Plasma Flows. In: Proceedings of 5th European Workshop on Thermal Protection Systems and Hot Structures, Nordwijk, Netherlands, ESA SP-631 (2006)
11. Lezuo, M.K.: Wärmetransport in H2-transpirativ gekühlten Brennkammerkomponenten. Ph.D. thesis, RWTH Aachen, Germany (1998)
12. Luikov, A.V.: Heat and Mass Transfer with Transpired Cooling. International Journal of Heat and Mass Transfer 6, 559–570 (1963)
13. Rannie, W.D.: A Simplified Theory of Porous Wall Cooling, JPL California Institute of Technology, Pasadena, USA (1947)

14. Rodet, J.C., et al.: Etude en souflerie thermique du refroidissement de parois poreuses par effusion de gaz. Revue Generale de Thermique 37, 123–136 (1997)
15. Rubesin, M.W.: An Analytical Estimation of the Effect of Transpiration Cooling on the Heat-Transfer and Skin-Friction Characteristics of a Compressible, Turbulent Boundary Layer. NACA TN 3341 (1954)
16. Rubesin, M.W., Pappas, C.C.: An Analysis of the Turbulent Boundary-Layer Characteristics on a Flat Plate with Distributed Light-Gas Injection. NACA TN 4149 (1958)
17. Schmücker, M., Grafmüller, A., Schneider, H.: Mesostructure of WHIPOX all Oxide CMCs. Composites, Part A 34, 613–622 (2003)
18. Serbest, E.: Untersuchungen zur Anwendung der Effusionskühlung bei Raketenbrennkammern. Ph.D. thesis, RWTH Aachen, Germany (2002)
19. Sucec, J., Oljaca, M.: Calculation of Turbulent Boundary Layers with Transpiration and Pressure Gradient Effects. International Journal of Heat and Mass Transfer 38, 2855–2862 (1995)

Qualification of Active Cooling Concepts in Ground Facilities

B. Esser and A. Gülhan

Deutsches Zentrum für Luft- und Raumfahrt (DLR)
Abteilung Windkanäle
des Instituts für Aerodynamik und Strömungstechnik
Linder Höhe, D-51147 Köln, Germany
burkard.esser@dlr.de

Summary

Two active cooling concepts were tested concerning their capabilities with respect to surface cooling in hypersonic flow. Film cooling was analysed in cold hypersonic flow at realistic Reynolds numbers. Cooling efficiency was found highly sensitive to injection geometry and flow inclination. Best performance was observed for helium with a heat flux reduction up to 90% compared to the non-cooled case. Transpiration cooling was thermally qualified in high enthalpy flow. The coolant was supplied through a porous material integrated in the surface. Substantial cooling could be achieved in the injection area and further downstream at low coolant flow rates. Optimal flow rates could be identified for several coolants.

1 Introduction

Reliable qualification of key technologies for future space vehicles, as e.g. improved cooling concepts basing on new material developments, is not possible without dedicated experiments in ground test facilities. Although the capabilities of numerical simulations have significantly improved in the past decades, the thermo-mechanical behaviour of highly loaded components of space vehicles during the hot phase of atmospheric entry can be predicted with larger restraints only, even for state-of-the-art technologies. New developments need additional experimental investigation in test facilities that are able to establish high enthalpy hypersonic flow conditions for time periods that are characteristic for atmospheric entry, i.e. at least several minutes. These requirements are met by the LBK facility which is one of the European key facilities for qualification of thermal protection systems. In LBK realistic combinations of convective and radiative heat loads can be applied and components can be tested in an atmosphere with a realistic gas composition at realistic Mach numbers. Other parameters, as e.g. Reynolds numbers, binary scaling factors and boundary layer thickness, cannot be duplicated in such facilities, in general. Therefore, a combined investigation with other ground test facilities is required to match all important parameters. Possible complementary facilities are short-duration high enthalpy facilities, as e.g. shock tunnels, that provide realistic Reynolds and Mach numbers, or cold hypersonic facilities, as e.g. the hypersonic windtunnel H2K,

A. Gülhan (Ed.): RESPACE – Key Tech. for Reusable Space Sys., NNFM 98, pp. 104–131, 2008.
springerlink.com

where boundary layer transition can be simulated at realistic Reynolds numbers, but on a low surface temperature level.

Two separate investigations were performed that represent two stages of the above mentioned combined ground testing philosophy. The first activity was strongly linked to cooling concept that is described in section 6. It focused on the thermal qualification of transpiration cooling in the arc heated facility LBK. All parameters that transpiration cooling is presumed to depend on, i.e. porous material, coolant gas, and coolant mass flow rate, were systematically varied in order to identify their influence on cooling efficiency. The second activity was carried out in the cold hypersonic windtunnel H2K and was linked to the investigations on film cooling concepts that are presented in sections 8 and 9. Details of this activity are reported in section 2, the qualification of transpiration cooling will follow in section 3.

2 Film Cooling in Cold Hypersonic Flow

Film cooling has been found to be one of the most efficient measures to protect propulsion components from high aerothermal loads [6, 7, 8, 12, 15]. But its application for external aircraft components, in particular in hypersonic flows, has still to be developed and demonstrated. Therefore, in the frame of this work the feasibility of film cooling for the protection of external hot structures in the hypersonic flow regime was investigated. A wedge model with variable injection geometry and cooling medium was used to perform a systematic experimental study in DLR's hypersonic wind tunnel H2K. In addition to feasibility information the experiments were expected to provide a better understanding of film cooling physics for a hypersonic flight environment.

Infrared thermography was used to measure the temperature distribution on the model surface. By using a material with a low thermal conductivity for the principal parts of the model surface the heat flux distribution could be determined from the measured surface temperature history. In combination with Schlieren pictures these measurements were used to interpret the boundary layer flow properties, the influence of free stream flow parameters, different coolant injection geometries, and flow parameters of the coolant.

The tests were performed at a Mach number of 6 and two different Reynolds numbers. The variation of the Reynolds number should allow to identify the conditions for a possible undesired boundary layer transition.

2.1 Wedge Model

A wedge model consisting of a nose with 45° wedge angle and a downstream flat plate as shown in Fig. 1 was used for the wind tunnel experiments. In order to study different parameters the model was designed in a modular way. It consists of three main components: a wedge type nose, a coolant injection slot and the main body with a flat plate on the flow side. The main body was mounted to the H2K model support system by screws.

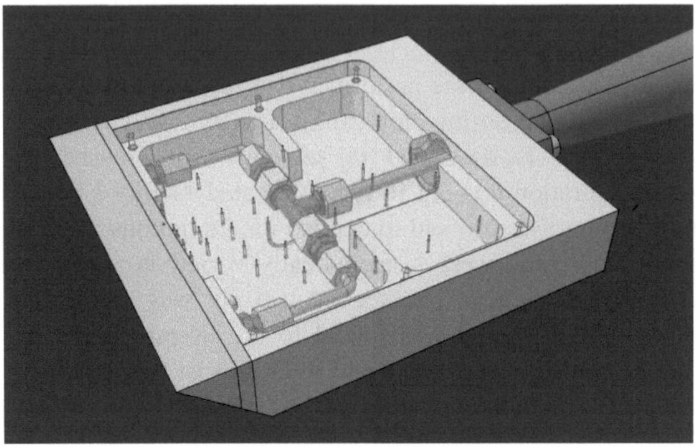

Fig. 1. 3D-view of the H2K wedge model

The nose part and the injection slot could be changed. Two different noses were available, one with a sharp edge and a blunt nose with a radius of 5 mm. The wedge angle was 45° for both. Variation of the nose part allowed to check the influence of the boundary layer thickness and entropy layer on the results. Different injection slots could be placed between the nose and the flat plate part. They were connected to a coolant reservoir inside the model's main body. The reservoir was equipped with a pressure and a temperature sensor.

The flat plate on top of the main body was made of PEEK material. Due to its emissivity value of 0.95 and its low thermal conductivity PEEK is well suited for heat flux determination from infrared spectroscopy. It is applicable at temperatures between –65°C and 250°C. Its thermal properties at room temperature are listed in Table 1. The properties change with temperature and a linear correlation was applied for the determination of heat flux rates from measured surface temperatures.

Table 1. Material properties of PEEK at 20°C

property	
Surface emissivity [–]	0.95
Thermal conductivity [W/(mK)]	0.25
Density [kg/m^3]	1320
Specific heat [J/(kgK)]	1656

2.2 Test Parameters

The main test parameters were the free stream conditions, model geometry, coolant gas and coolant injection. At a common Mach number of 6 two free stream conditions with a significantly different Reynolds number were applied. The main parameters are listed in Table 2. Furthermore, tests were performed at two angles of attack, i.e. $\alpha = 0°$ and $\alpha = 20°$. The corresponding model coordinate system is shown in Fig. 2.

Table 2. Free stream parameters

Parameter	FC-A	FC-B
Reynolds number [1/m]	$2.8 \cdot 10^6$	$14.3 \cdot 10^6$
Total temperature [K]	730	470
Reservoir pressure [bar]	6.0	16.5
Flow rate (free stream) [kg/(m^2 s)]	16.85	57.75
Flow rate (post-shock, $\alpha = -20°$) [kg/(m^2 s)]	55.56	190.4

Air, argon and helium were used as coolants. They were chosen in order to achieve sufficient difference in heat capacity, thermal conductivity, and density. The main properties of the coolant gases are listed in Table 3.

Table 3. Thermodynamic properties of coolants

Thermodynamic property	Air	Helium	Argon
Molar mass [kg/kmol]	28.96	4.003	39.95
Density (at standard conditions) [kg/m^3]	1.2929	0.1785	1.784
Specific heat at constant pressure [J/(kgK)]	1007	5238	524
Specific heat at constant volume [J/(kgK)]	720	3214	323
Ratio of specific heats [–]	1.40	1.63	1.62
Gas constant [J/(kgK)]	287	2024	201
Thermal conductivity [10^{-3} W/(mK)]	25.6	148	17,3
Dynamic viscosity [10^{-6} kg/(ms)]	18.19	19.6	22.11
Prandtl number [–]	0.72	0.69	0.66

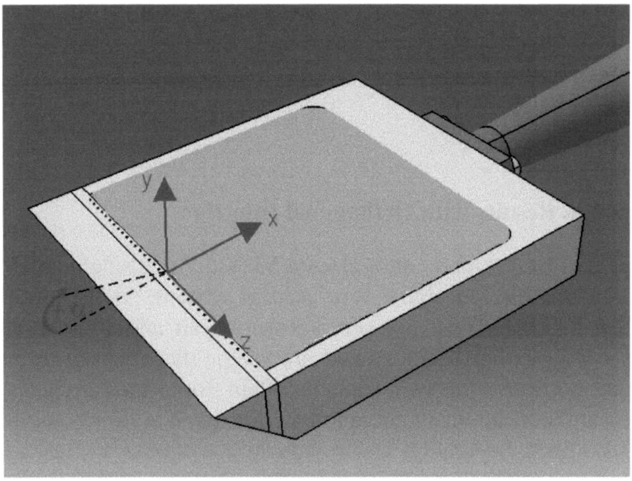

Fig. 2. Model with definition of coordinate system

The coolant mass flow rate could be varied between 0 and 0.8 g/s. For the comparison of cooling efficiency the blowing ratio

$$F = \frac{(\rho\, u)_{coolant}}{(\rho\, u)_e} \tag{1}$$

was used which relates the coolant flow rate

$$(\rho\, u)_{coolant} = \dot{m}_{coolant} \big/ A_{injection} \tag{2}$$

to the external flow rates at the boundary layer edge which are included in Table 2 for the two angles of attack.

The geometry of coolant injection was an important test parameter. Due to the modular model design it could be changed easily by changing the injection slot. Six injection slots were used during the test campaign. The corresponding hole configurations are listed in Table 4 The diameter of holes was either 0.5 mm or 1 mm.

Single hole injection was used for orthogonal injection only. Slots with multiple holes aligned in a row with a distance of 5 mm between the holes were used for orthogonal injection, for oblique injection at an angle of 30° to the surface and for tangential injection. For slots R3 and R4 the holes were placed on one half of the slot area.

Table 4. Injection slots

Name	Type	Number of holes	Diameter of holes [mm]	Injection angle
S1	Single hole	1	1.0	90°
S3	Single holes	3	1 x 0.5 and 2 x 1.0	90°
R1	Row of holes	35	1.0	90°
R2	Row of holes	34	0.5	90°
R3	Row of holes	17	0.5	30°
R4	Row of holes	17	0.5	0°

2.3 Experimental Results with Orthogonal Injection

First tests were carried out on a configuration S3 with three single holes, one 0.5 mm wide and two 1 mm. Air and argon were used as coolant. As the infrared images in Fig. 3 show, the thermal effects are very similar for air and argon at a coolant flow rate of 0.01 g/s. Distinct horizontal streaks are visible that emanate from the injection hole. The streaks correspond to vortices in the main flow which are generated by local flow separation at the locations of coolant injection, even at flow condition A with the low Reynolds number. Separation and re-attachment shocks as well as re-attaching vortices lead to a strong local aerothermal heating [2, 9].

The effect is more intense for 1 mm holes, but for the 0.5 mm hole, which is located in the lower part of the IR image, it is still obvious. Therefore, injection configuration S3 causes local heating instead of cooling.

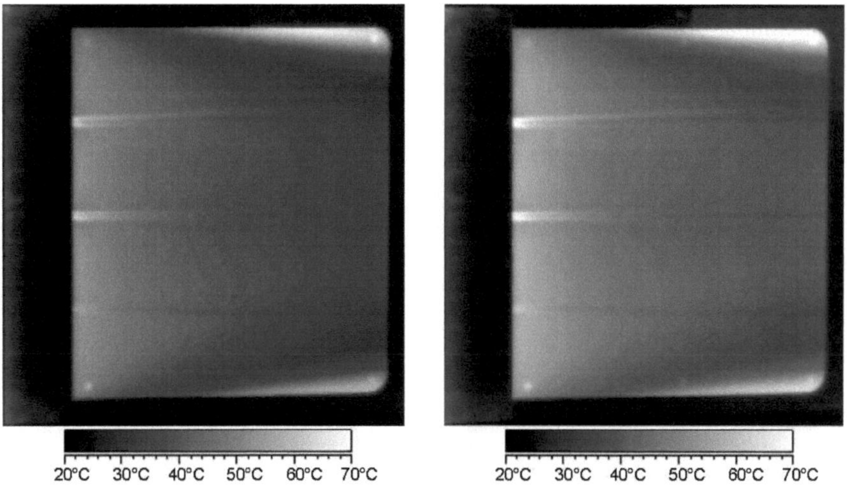

(a) air (b) argon

Fig. 3. Air and argon injection at 0.01 g/s (F = 0.34) through single holes with 0.5 mm and 1 mm diameter

Injection slot R2 (34 holes of 0.5 mm) causes similar heating phenomena at a total coolant mass flow rate of 0.02 g/s, although the flow rate per hole is significantly lower. The corresponding infrared images are shown in Fig. 4. Compared to air and argon the aerothermal heating appears more intense for helium cooling indicating a more significant disturbance of the main flow for this gas. Helium has a significantly lower density compared to the other two gases. Therefore, its volume is much higher causing a stronger disturbance to the laminar boundary layer.

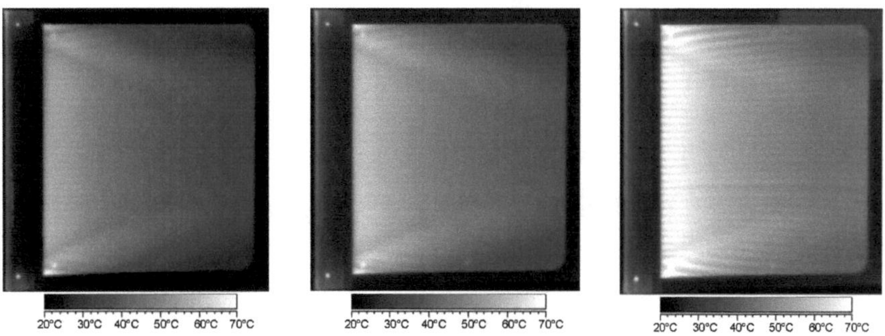

(a) air (b) argon (c) helium

Fig. 4. Air, argon and helium injection at 0.02 g/s (F = 0.18) through a row of 34 holes with 0.5 mm

Aerothermal heating increases with larger coolant mass flow rates. For single hole injection with slot S1 the heating caused by re-attachment of separated flow and vortices becomes stronger for the high coolant flow rates ($F = 5.74$) as shown in

Fig. 5. Similar results were obtained for slot R2. The infrared images in Fig. 6 indicate a stronger aerothermal heating when the coolant mass flow rate is increased by a factor of two.

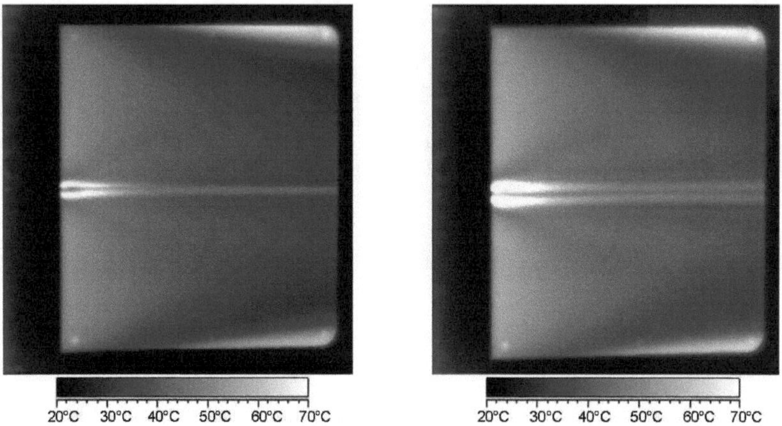

(a) $F = 2.04$ (b) $F = 5.74$

Fig. 5. Cold air injection with slot S1 at flow rates of 0.027 g/s (F = 2.04) and 0.076 g/s (F = 5.74)

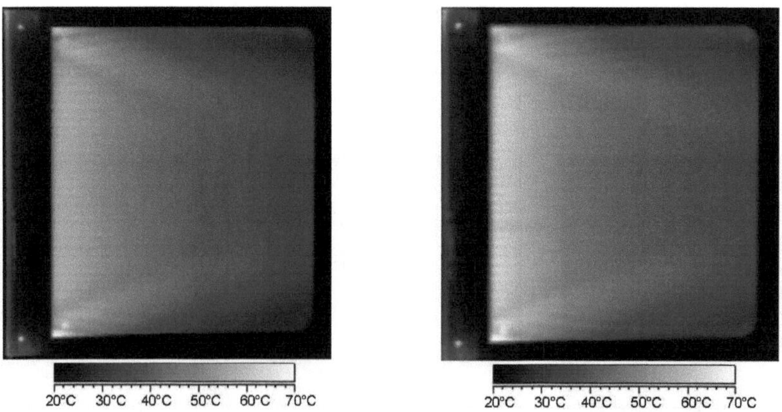

(a) $F = 0.18$ (b) $F = 0.36$

Fig. 6. Cold air injection with slot R2 at different flow rates of 0.02 g/s (F = 0.18) and 0.041 g/s (F = 0.36)

2.4 Experimental Results with Tangential Injection

All tests with orthogonal injection showed that the boundary layer is disturbed by the coolant causing aerothermal heating instead of cooling. Since the situation could not significantly be improved by reducing the injection angle from 90° to 30°, it was decided to change injection to tangential injection with slot R4. It was realized by mounting a deflection plate above the injection part for orthogonal injection.

In order to see the effect of cooling instantaneously only one half of the injection slots was used for the cold gas injection. The infrared images in Fig. 7 show that the lower part is not cooled and could be used as reference. The figures correspond to different coolant mass flow rates. At the lowest mass flow rate of 0.039 g/s ($F = 0.69$) cooling is observed in a very narrow area close to the injection slot. Further downstream there is no significant difference between the cooled part and the reference surface.

Increasing the coolant mass flow rate to 0.187 g/s the cooled surface area downstream the injection increases. Except for the lateral boundaries of the cooled area no heating caused by the vortices or boundary layer tripping has been noticed. At the highest mass flow rate of 0.383 g/s cooling is strong as well close to the injection. But further downstream the pattern becomes more streaky indicating a increasing heat transfer to the surface increases due to vortex attachment and boundary layer tripping. At one location of the reference model the surface is heated due to similar effects downstream of a leak in the injection slot.

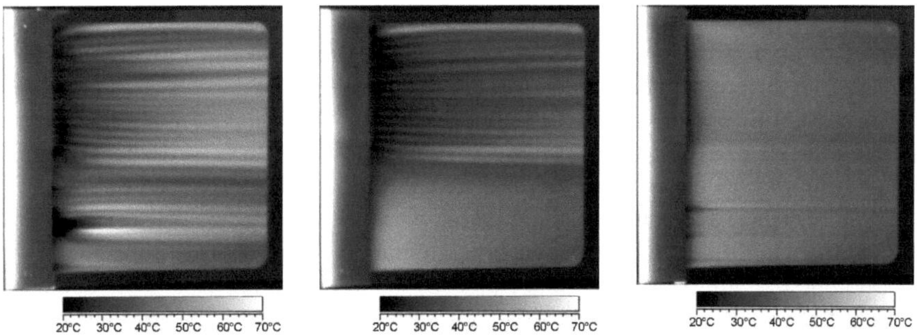

(a) $F = 6.81$ (b) $F = 3.32$ (c) $F = 0.69$

Fig. 7. IR images of tests at $\alpha = 0°$ with air cooling at injection rates of 0.383 g/s (F = 6.81), 0.187 g/s (F = 3.32), 0.039 g/s (F = 0.69)

A comparative technique is applied for the analysis of the test results. On the cooled side, Stanton numbers are evaluated along a line oriented in flow direction. Afterwards, the Stanton number distribution is determined for the corresponding line on the reference side. So, the ratio between the Stanton numbers indicates the cooling efficiency. By this procedure influences from geometry and free stream properties are eliminated. Cooling is effective as long as the ratio between cooled and non-cooled Stanton number remains below unity. A ratio above 1 corresponds to ineffective cooling, since in this case the heat on the cooled side is larger than the non-cooled reference.

In Fig. 8 Stanton number ratios are compared for three tests with air as coolant at different mass flow rates. At the lowest mass flow rate of 0.039 g/s cooling is restricted to the immediate vicinity of the injection area as it had already been

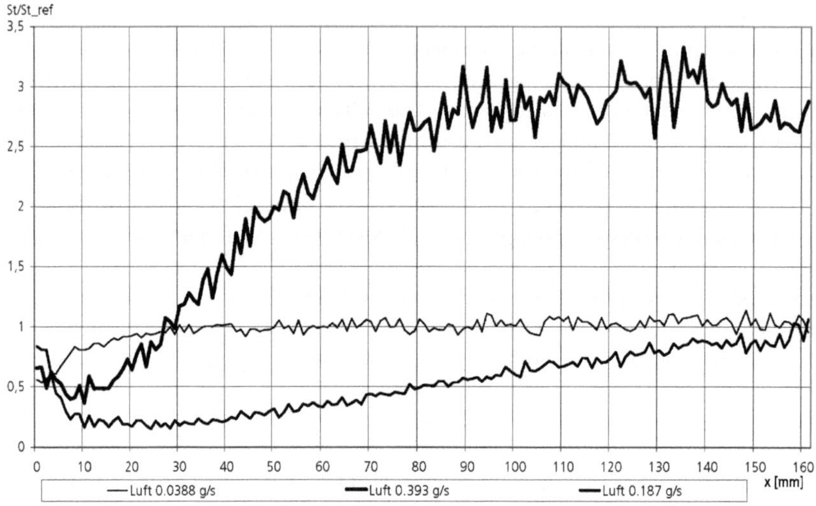

Fig. 8. Stanton number ratios in axial direction for different injection rates of air

observed from Fig. 7. From a location 20 mm downstream the ratio is close to 1, indicating no difference between cooled and reference side.

At a coolant flow rate of 0.187 g/s effective cooling is achieved along the complete length of the PEEK plate. Best cooling is provided at distances between 10 mm and 30 mm from the injection. Here the heat flux to the surface is reduced to 20% of the reference. Further downstream, the Stanton number ratio gradually increases. The efficiency of cooling decreases due to a disintegration of the coolant film.

Doubling the injection rate to a value of 0.383 g/s gives rise to a completely different behaviour. Due to increased shear stress the coolant film becomes unstable tripping the boundary layer. Cooling is observed within the first 30 mm only. Further downstream the heat load to the surface is increased significantly, partially reaching up to 300% of the non-cooled value.

At an injection rate of about 0.19 g/s, where the most effective cooling was observed in air, tests were performed with argon and helium as well. In Fig. 9 the Stanton number ratios of all three coolants are compared. The general behaviour is similar. Effective cooling is observed along the complete length. Maximal efficiency is achieved close to the injection, further downstream the efficiency gradually decreases.

Along the first 20 mm there is no remarkable difference between the coolants. In that region, cooling is dominated by model geometry. The backward facing step at the coolant injection generates a detached flow with a recirculation near the surface that reduces heat flux. Differences between the particular coolants become evident further downstream. Compared to air, argon behaves almost identical along the first 40 mm with an identical minimum Stanton number ratio of 0.2. At larger distances to the coolant injection argon cooling is more efficient. At the end of the PEEK plate, where the potential of air cooling has exhausted, argon still provides a Stanton number ratio of 0.7.

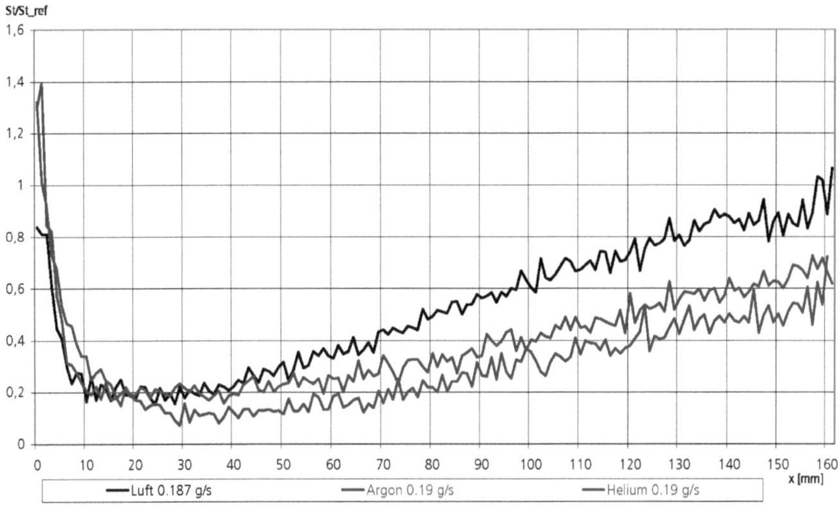

Fig. 9. Comparison of Stanton number ratios along an axial line for air, argon and helium cooling

Best performance, however, is obtained for helium cooling. It provides a reduction of the Stanton number to about 10% of the non-cooled reference and the efficiency remains best up to the end of the PEEK plate.

Differences between the coolants become evident in the infrared images as well (see Fig. 10). Compared to air cooling the surface temperature at the coolant injection is significantly lower with argon. In addition, the vortices at the sides of the cooled regime are more evident in the surface temperature distribution.

For helium the temperature difference between coolant injection and the end of the PEEK plate is lowest. In addition, the streak pattern on the cooled side is not as

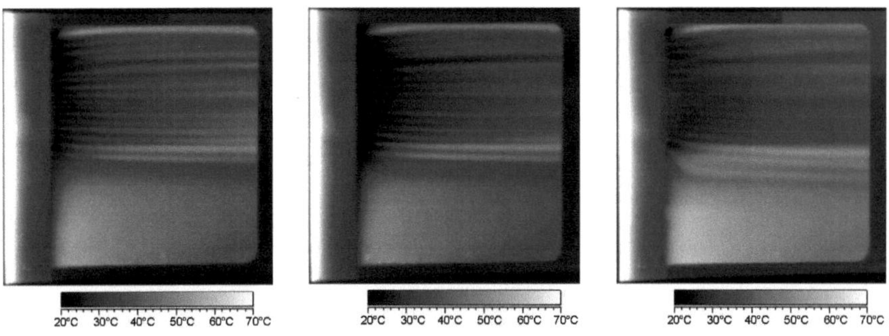

(a) air (b) argon (c) helium

Fig. 10. IR images of tests with different coolants at 0.19 g/s (F = 3.37) pronounced as it is observed for air and argon, where individual coolant jets can be distinguished much better. There is a penetration of coolant into the reference side for all three coolants, but helium seems to penetrate farther than air and argon.

The better performance of helium can be explained by its heat capacity, which is about five times higher compared to air and 10 times higher compared to argon. But since argon performes better than air, heat capacity is not the only parameter that takes influence on cooling efficiency.

When the model is inclined to the flow by 20°, the effect of air cooling remains almost negligible at the lowest mass flow rate of 0.039 g/s. Just the recirculation area downstream of the coolant injection is more prominent than at $\alpha = 0°$ as comes out from the infrared image in Fig. 11. Streaks of individual coolant jets can be guessed

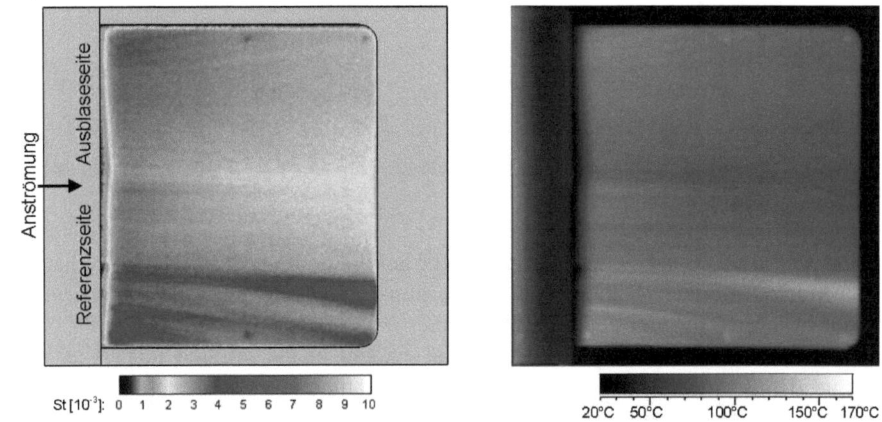

(a) Stanton number (b) Temperature

Fig. 11. Stanton number and surface temperature distribution for air cooling at 0.039 g/s (F = 0.21) and $\alpha = -20°$

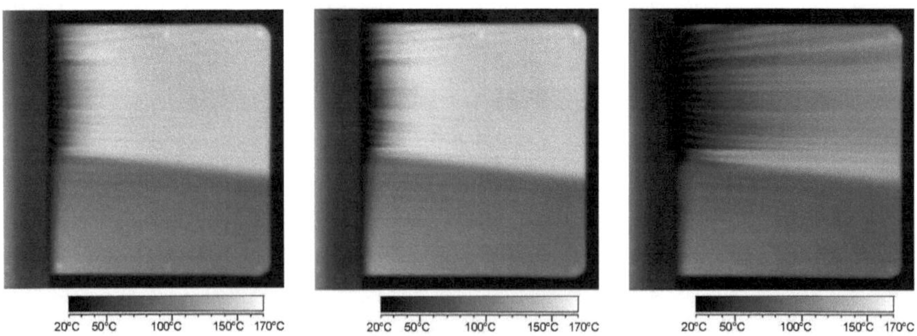

(a) air (b) argon (c) helium

Fig. 12. IR images of the tests at $\alpha = -20°$ with 0.184 g/s air, argon and helium cooling (F = 0.99) only on the cooled side of the model. The clear streak on the non-cooled reference side refers to an unintended leakage.

Situation changes significantly when the injection rate is increased to 0.184 g/s. As comes out from the infrared images shown in Fig. 12 the recirculation area is enlarged compared to Fig. 11 due to the higher amount of injected gas. But at the downstream end of the recirculation area the boundary layer is tripped and the surface temperature is increased by about 50% compared to the non-cooled reference side.

Only minor differences are observed between air and argon. Downstream of the cool injection area there is a zone with a streaky flow pattern corresponding to boundary layer transition. Further downstream the temperature distribution is nearly homogeneous at an increased level.

The characteristics from the infrared images are confirmed by the Stanton number ratios shown in Fig. 13 which were evaluated along an axial line. The results for air and argon are almost identical, and cooling is obtained for the first 20 mm only. The minimum value is close to 0.2 which is comparable to the value at $\alpha = 0°$. In the transition regime which extends up to $x = 50$ mm the Stanton number ratio increases to a value of 2. Except for a minor increase along the last 30 mm, this level is kept until the end of the PEEK plate.

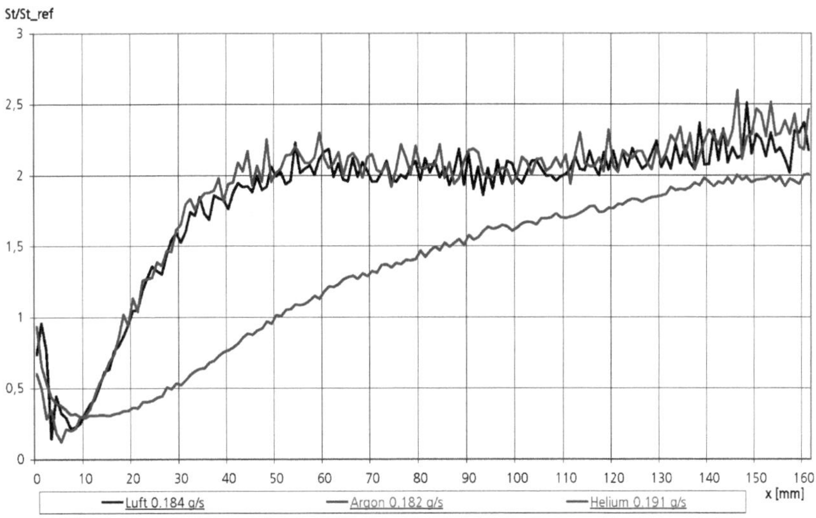

Fig. 13. Comparison of Stanton number ratios along an axial line for air, argon and helium cooling at α = -20° (F = 0.99)

At first glance, the situation is different for helium cooling. The infrared image shows a significantly larger cooling zone close to the injection. The Stanton number ratio remains below 1 for a section of 50 mm compared to 20 mm for air and argon. The temperature gradient inside the transition regime is smaller and the transition regime extends almost to the end of the plate. Along the plate's last 20 mm the Stanton number ratio is nearly constant at a value of 2. So, for all three gases the same

Stanton number ratio is achieved at the end of the transition zone. This gives rise to the assumption that the Stanton numbers ratios follow a common trend, but with a different characteristic length scale for helium compared to air and argon. The different length scale might be related to the sound velocity, which is almost identical for air and argon and more than three times larger for helium. Due to sonic injection the initial coolant velocity is significantly higher for helium which enables a deeper penetration into the boundary layer.

2.5 Concluding Remarks

An experimental investigation was performed on a flat plate model with respect to film cooling in the laminar hypersonic flow field of H2K at a free stream Mach number of 6. Model geometry, injection angle, coolant gas and free stream conditions were varied to find out their potential concerning the reduction of aerothermodynamical heat loads. An infrared camera system was applied to measure surface temperature distributions on the model and the corresponding heat loads were evaluated from the temperature evolution.

No cooling effect could be achieved when the coolant was injected perpendicular to the flow. Air, argon, and helium were applied as coolants, but no heat reduction was observed, neither for a single injection nor for multiple injections laterally aligned in a row. Instead of cooling locally enhanced heat loads were observed which increased with increasing coolant mass flow rate. Due to the orthogonal injection the coolant jets acted as obstacles in the hypersonic flow. At the injection area strong horseshoe-like vortices develop that are known to cause increased heat loads.

The situation could not be improved by reducing the injection angle from 90° to 30°. At substantial coolant mass flow rates the same behaviour was observed, weak cooling could be achieved for very low mass flow rates only.

Efficient cooling was obtained with tangential injection at the bottom of a backward facing step. At zero angle of attack all three coolants provided a substantial reduction of heat fluxes for a blowing ratio of 3.3 Best performance was observed for helium, which reduced the heat flux up to 90% compared to the non-cooled case. The maximum local heat flux reduction was 80% for air and argon. Since the cooled area extended further downstream, the global performance was slightly better for argon.

When the model was inclined to the flow by 20° the boundary layer got tripped for all coolants at substantial coolant mass flow rates with a blowing ration close to 1. The heat flux to the surface was enhanced compared to the non-cooled reference case. Heat enhancement was strong for air and argon and more moderate for helium.

3 High Enthalpy Qualification of Transpiration Cooled TPS

Thermal qualification of structures that are exposed to extreme thermal loads, e.g. during Earth re-entry or planetary entry, needs long-duration ground test facilities. During atmospheric entry thermal loading is a combination of convective and radiative heat loads. Due to the non-linear behaviour of radiation the scenario does not allow for thermal scaling, but needs testing at real temperature levels.

Transpiration cooling is assumed to be mainly influenced by the porous surface material, the coolant used, and the coolant's mass flow rate. These parameters were varied during the qualification test campaign in order to find their influence on cooling efficiency.

3.1 Test Facility

The tests on transpiration cooling were performed in DLR's arc heated facility LBK, which consists of the two tests legs L2K and L3K. The facility is one of the European key facilities for the qualification of thermal protection systems and can provide sufficient testing times and realistic atmospheric conditions for planetary entry. A sketch of the facility is plotted in Fig. 14, details can be found in [4, 3].

Central part of each test leg is an arc heater. At L2K, there is a Huels-type heater with an electrical power of 1 MW and at L3K a 6 MW segmented arc heater. Downstream of the arc heater the flow is accelerated to hypersonic velocities by conical nozzles with a half angle of 12°. Depending on the nozzle's exit diameter Mach numbers in the range from 4 to 10 can be achieved at stagnation pressures up to 350 hPa.

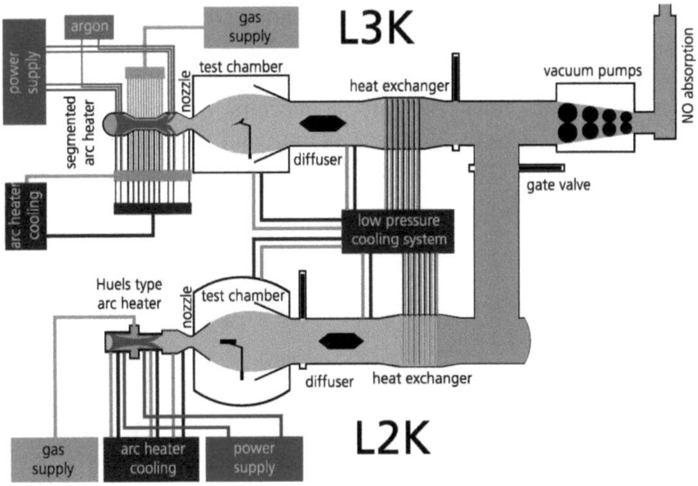

Fig. 14. Sketch of the LBK facility

During the project both tests legs have been used. First screening tests were run in L2K in order to check the applicability of transpiration cooling and the operation of the cooling supply systems. Afterwards testing was switched to the L3K facility, where higher enthalpy levels, surface temperatures and surface pressures can be achieved.

3.2 Model and Instrumentation

Thermal testing of transpiration cooled structures in a hypersonic high enthalpy flow field requires an experimental setup that allows for continuous coolant supply at adjustable mass flow rates through a porous sample in the surface of windtunnel model. For reliable operation the porous sample was directly connected to a coolant reservoir inside the model. The coolant itself was supplied from K-bottles outside the test chamber with a controller installed in the supply line. This concept was checked in two screening test campaigns in L2K and L3K which are described in more detail in section 6. The general concept was approved by the screening tests, only details of the reservoir design had to be modified, e.g. adhesives were identified unsuitable for the connection between porous sample and reservoir.

For the final test campaign in the L3K facility a flat plate model was used as shown in Fig. 15. The porous sample was integrated on the model's left hand side (viewed in flow direction). Square 60 mm wide porous samples with a thickness of 6 mm were used. The samples were embedded in a plate of carbon reinforced silicon carbide (C/C-SiC) which is a well qualified reusable high-temperature thermal protection material. Detailed information on the C/C-SiC material is given in [5, 10]. Except for the porous sample the model is symmetric with respect to its mid plane in flow direction. The two sides differ by the porous sample only. Since the flow field is symmetric to the same plane as well, the non-cooled side could be used as reference.

The complete setup was installed on top of a water-cooled model holder with a blunt nose, which is water-cooled as well. The cubic coolant reservoir made of C/C-SiC was mounted just on top of the model holder's base plate. There are three connections to the reservoir's side wall (see Fig. 15a). The left one corresponds to the coolant supply line, the other two are connectors for measuring pressure and temperature inside the reservoir. Nitrogen, helium and argon were used as coolants, the coolant's mass flow rate could be set to 0.2-10 g/s.

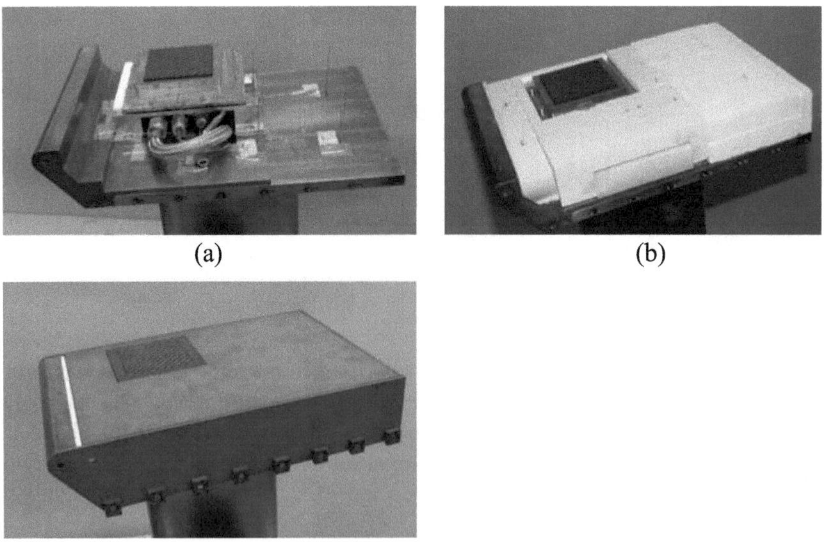

(a) (b) (c)

Fig. 15. Model for transpiration cooling testing

The porous sample ws installed on top of the reservoir. A tight connection was obtained by a carbon seal which was compressed by a jacket frame made of high-temperature alloy PM2000. The remaining space between the C/C-SiC plate on top and the base plate was filled with insulating material. The insulation prevented the aerothermal heating on the top from being influenced by the cooled parts of the model holder. Furthermore, the insulation material avoided internal radiation, which might lead to a substantial heating to the model's interior due to the expected high surface temperatures.

Three different porous sample materials were tested. Although all of them made of carbon reinforced carbon (C/C), there were differences with respect to porosity, permeability, and structural details. The majority of tests was peformed with samples made of "standard C/C" having a porosity of about 17%. This material is referenced as standard, since it had been successfully applied to transpiration cooling in a different environment, i.e. a combustion chamber [11, 14]. The "highly porous C/C" is similar to standard C/C, but with a porosity of 44%. Standard C/C and highly porous C/C have the carbon fibres oriented parallel to the top surface, while for the third type, "stratified C/C", the layers are oriented perpendicular to the surface. The main properties of all samples that were tested are listed in Table 5.

Table 5. Tested samples

Sample	Material	Porosity [%]
S1	Standard C/C	17
S2	Standard C/C	17
S3	Row of holes	17
H1	High porosity C/C	44
P1	Stratified C/C	16

The model was primarily instrumented for temperature measurements. In total 17 thermocouples were installed in the interior, 12 either directly below the C/C-SiC cover plate or inside the insulation material. The coolant's state in the reservoir was monitored during tests by a thermocouple and a pressure gauge. The remaining thermocouples were used to monitor the temperatures of reservoir walls, sample and base plate. The locations of the thermocouples are plotted in Fig. 16.

In addition to the temperature measurements inside the model the surface temperature distribution was measured by infrared cameras. During all tests a camera with a high temperature measuring range between 350°C and 2000°C was used. The temperature range was well-suited for the non-cooled parts of the surface. For the porous probe, however, the camera was working near to or even below its low temperature limit. In order to obtain a better temperature of this particular area a second infrared camera was added which was set for measurements in the range of 0–500°C. The camera measurements were supported by pyrometer measurements at specific spots on the surface, i.e. the centre of the porous sample, the corresponding location on the non-cooled side, and a location upstream of the sample. Since the pyrometers work in the near infrared range close to 1 µm, their measurements are almost not affected by uncertainties in the surface material's emissivity. Therefore

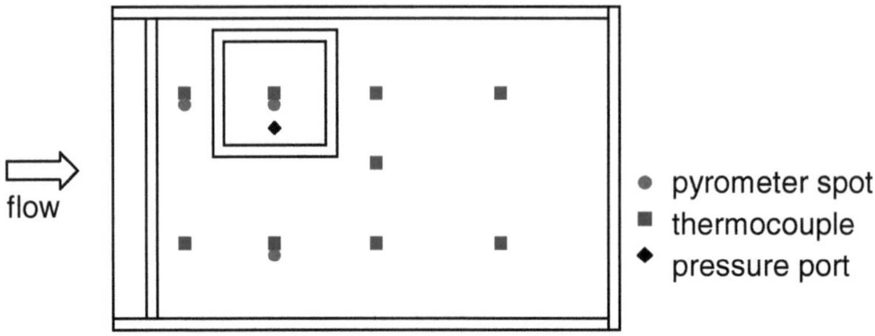

Fig. 16. Pyrometer spots and locations of intrusive measurements

they could be used to correct the results of the infrared cameras which operate in the far infrared regime with regard to surface emissivity [1].

3.3 Flow Condition and Test Parameters

A nozzle with a throat diameter of 29 mm and an exit diameter of 300 mm was used for the tests. Air was used as working gas. Based on the facility's operating conditions the nozzle flow field was computed using the CEVCATS-N code [13]. The computation includes the effects of non-equilibrium thermodynamics and chemistry. The resulting Mach number distribution along the nozzle is plotted in Fig. 17. The flow Mach number increases almost uniformly along the nozzle. In lateral direction the flow field is nearly homogeneous providing constant flow properties along the width of the model which is placed 300 mm downstream of the nozzle exit area. The flow properties that were obtained at model location are listed in Table 6 together with the operating conditions.

All tests were performed according to an identical test procedure. After ignition of the arc heater the flow conditions were gradually changed during facility startup until the desired operating conditions had been achieved. During this period the model was

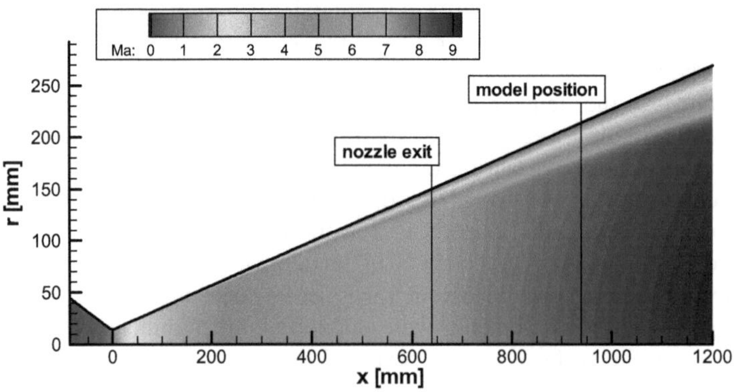

Fig. 17. L3K flow field

placed outside the flowfield in the background area of the test chamber in order to avoid the measurements from being influenced by changing flow conditions. During this period the transpiration cooling system was started as well.

As soon as the desired flow conditions were reached and the coolant was being supplied at the defined mass flow rate, the test was started by moving the model to the axis of the flow field. Here, the model remained for the specified test duration which was either 180 s or 240 s. After this period, the model was moved out of the flow field, and the flow was stopped. Data acquisition was continued for about five more minutes, since the cooling phase provides additional information about the heating of the model.

Table 6. Facility operating conditions and flow conditions at model location

Reservoir pressure	[hPa]	4700
Reservoir temperature	[K]	5650
Total enthalpy	[MJ/kg]	11.6
Mach number	[–]	7.5
Free stream pressure	[Pa]	56
Free stream temperature	[K]	530
Free stream velocity	[m/s]	3873
N_2 mass fraction	[–]	0.762
O_2 mass fraction	[–]	0.004
NO mass fraction	[–]	0.007
N mass fraction	[–]	$< 10^{-4}$
O mass fraction	[–]	0.227

3.4 Test Results

Transpiration cooling of a structure in a high-enthalpy, hypersonic flow is mainly influenced by the applied porous material, the coolant used, and the coolant's mass flow rate. The objective of the test campaign in L3K was to investigate the influence of these parameters on the thermal behaviour of the model for two different heat loads which were generated by varying the model's angle of attack between 20° and 30°. Due to the high number of test parameters "standard C/C" was defined as baseline material and nitrogen as baseline coolant. Accordingly, tests with different coolants were performed on "standard-C/C" samples and systematic variations of the coolant mass flow rate were carried out with nitrogen cooling.

Individual tests were performed at constant coolant mass flow rate. Therefore, in one test only a single combination of parameters could be tested. Proceeding that way, the facility must be able to reproduce identical test conditions during subsequent runs. Figure 18 shows the results of pyrometer and thermocouple measurements from three tests at an angle of attack of 20°. The measurements were taken on the non-cooled reference side at a location that corresponds to the centre of the porous sample. The testing time was 180 s for all three tests. This period is marked by a red background colour. The signals coincide well proving the required reproducibility.

At the end of the testing time the surface temperature which is measured by the pyrometer reached a steady state level of 1015°C. Since the pyrometer has a low temperature limit of 900°C, the measurement becomes physical when the surface temperature is exceeding this limit. During the first 55 seconds the pyrometer signal is set by the internal electronics to an artificial value. The thermocouple which is placed 3 mm below the surface starts rising almost immediately when the model is moved into the flow field. At the end of the test it indicates a nearly constant temperature of 955°C.

When the heat load is increased by changing the angle of attack to 30° the surface temperature reaches 1170°C after 180 seconds testing as shown in Fig. 19. Below the cover plate, the temperature increases to 1120°C as indicated by the thermocouple measurements. Again, the temperature difference across the cover plate is close to 50 K, which agrees to the thermal properties of the material. The agreement of both, pyrometer and thermocouples signals, show a good agreement proving the reproducibility of test conditions at the higher heat load as well.

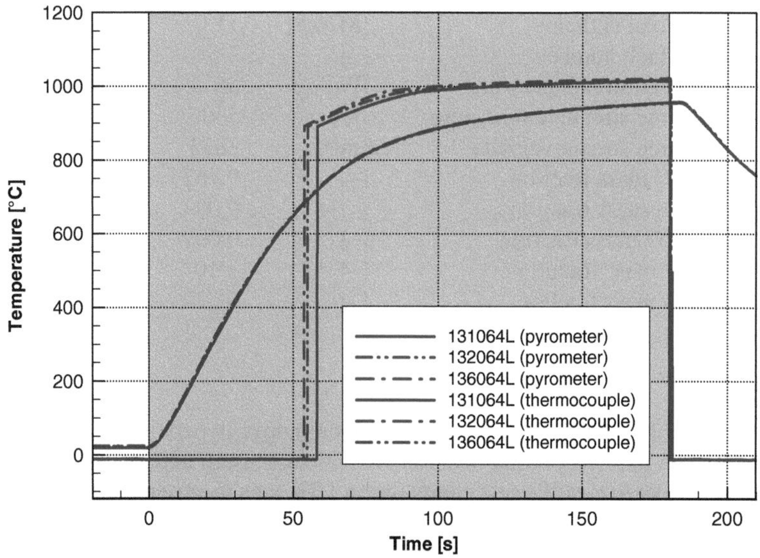

Fig. 18. Pyrometer and thermocouple measurements at the reference location from three tests at $\alpha = 20°$

The non-cooled side of the model also provides a kind of technological reference with regard to results of transpiration cooling, since the surface is completely made of reusable TPS material that is able to withstand the test condition without being cooled. When analysing transpiration cooling, however, with regard to its influencing parameters, e.g. coolant mass flow rate, the non-cooled side does not provide an ideal non-cooled reference, because the surface materials are different and their thermal and radiation properties as well, in general. Furthermore, the thickness of the surface layer is different as is the structure beneath. On the non-cooled side the model interior is filled with insulation, while there is open space below the porous sample. A more reliable non-cooled reference requires measurement on the porous sample itself.

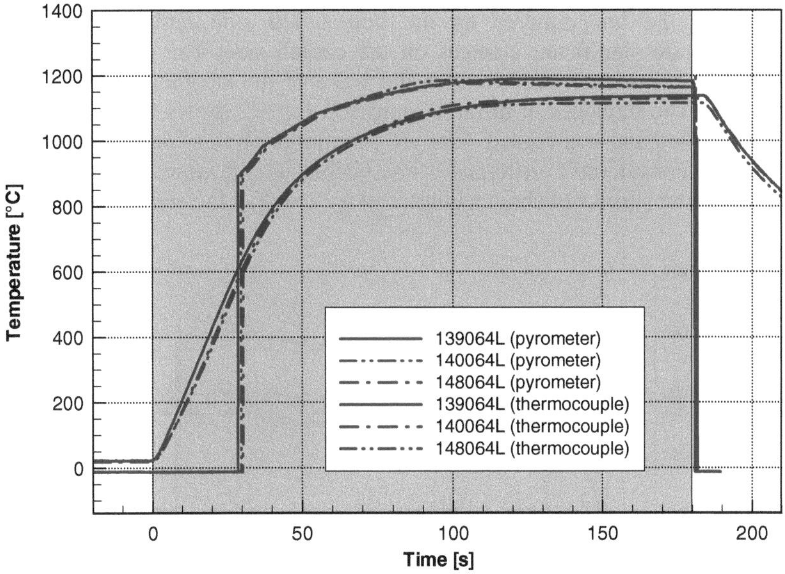

Fig. 19. Pyrometer and thermocouple measurements at the reference location from three tests at $\alpha = 30°$

Therefore, reference tests without coolant were included in the test matrix. Since the variation of coolant parameters was investigated on "Standard C/C", the same material was used for the non-cooled reference tests.

Figure 20 shows the results of temperature measurements on the top and rear side of the porous sample during the non-cooled reference test at a model inclination of 20°. The temperature on the top surface which is exposed to the flow has been taken from the infrared camera measurement, while the temperature on the bottom was measured by a thermocouple. At the end of the test the top surface is heated up to 980°C, which it by 35 K lower than the temperature on the non-cooled side of the model (see Fig. 18) due to the different materials' thermal properties. The temperature at the bottom reaches 732°C. The temperature difference between top and bottom side is significantly higher than measured for the cover plate which is mainly caused by the sample's thickness of 6 mm compared to a thickness of 3 mm for the cover plate.

Without coolant flow the sample's porosity is not beneficial, since it enables hot gas penetration into the reservoir. Therefore, the temperature inside the reservoir increases to 402°C at the end of the test, still increasing.

The surface temperature distribution is very homogeneous at the end of the reference test as comes out from the infrared image in Fig. 21. There are no temperature differences observed on the cover plate in lateral direction. Only the sample and its frame show slightly different temperatures, the sample due to its different emissivity value, the frame due to a different surface coating.

Figure 21b shows the same sample (Standard C/C) in a test with transpiration cooling. While the temperatures on the non-cooled side remain unchanged as expected, there are significant changes on the cooled side. The temperature of the sample itself is considerably reduced. In the wake of the sample, the cover plate is cooled as well. The sequence of infrared images in Fig. 22 shows that the differences between the cooled and non-cooled sides are even more obvious during the test. After 90 s of testing considerable differences are visible up the downstream end of the model. This region shrinks to about a quarter of its length at the end of the test.

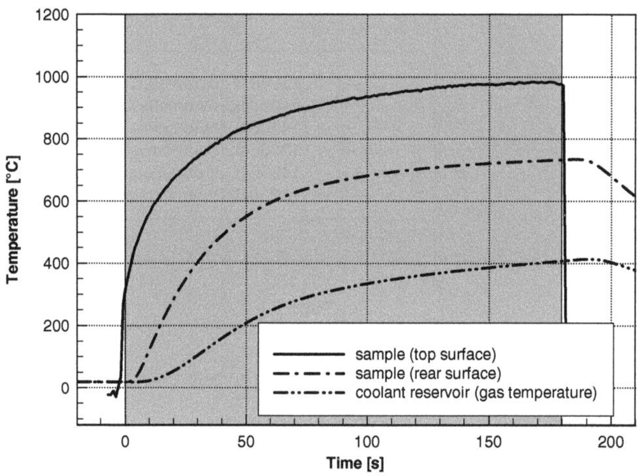

Fig. 20. Temperature measured on sample S1 and in the reservoir during the non-cooled reference test at $\alpha = 20°$

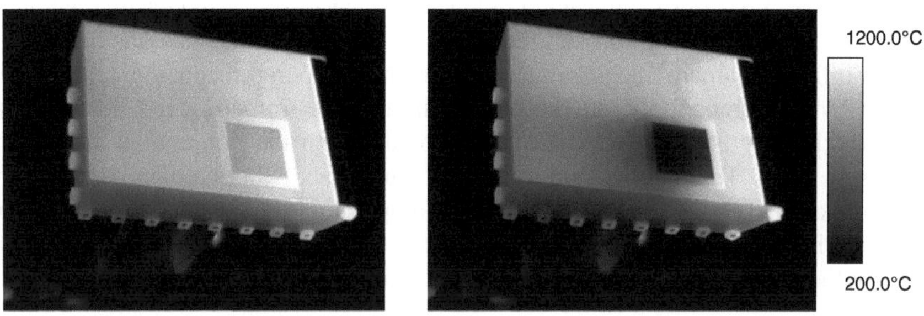

(a) non-cooled (b) cooled by 0.4 g/s N_2

Fig. 21. Infrared images at the end of a test with and without coolant flow (sample S1, $\alpha = 20°$, flow from right to left)

In the centre of the sample the temperature increases up to 280°C as shown in Fig. 23. Compared to the non-cooled reference test this is a reduction by about 700 K. Although the surface temperature did not reach a steady state condition after 180 s, the reduction is quite impressive. It becomes even more impressive when considering that a small part is caused by a gradual increase of the coolant's temperature.

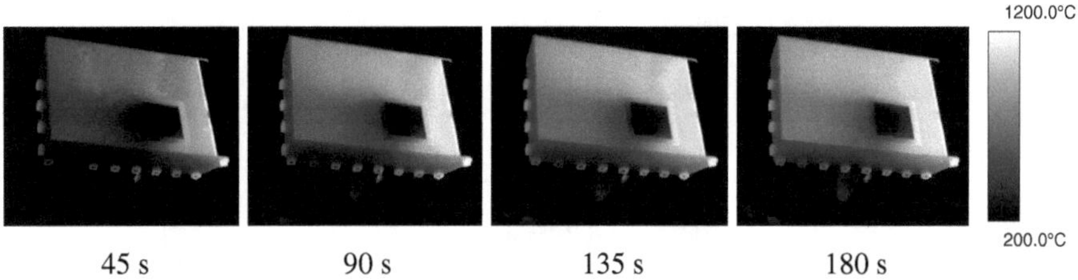

Fig. 22. Sequence of infrared images (sample S1, cooled by 0.4 g/s N_2, $\alpha = 20°$, flow from right to left)

During the test the coolant inside the reservoir heats up by about 65 K. The corresponding heat is mainly transferred along the side walls of the reservoir. It could hardly be avoided, since the reservoir had to be designed rigid with pressure-tight side walls and it had to be placed close to the hot surface.

When the model's inclination is increased to 30°, the observations are similar, but on a higher temperature level. Comparison of the infrared images in Fig. 24 from tests with and without coolant flow shows no differences in the temperature distribution on the reference side. For the non-cooled test the complete surface is homogeneously heated without obvious lateral temperature gradients. Again, cooling considerably affects the temperature evolution of the sample and the parts of the cover plate located downstream of the sample. As indicated by the sequence of infrared images in Fig. 25 the differences between the cooled and non-cooled parts are most significant after 45 and 90 seconds. Towards the end of the test the region with obvious differences becomes smaller. Due to the different coolant mass flow rates a direct comparison of the influence of transpiration cooling in the wake of the sample is not possible.

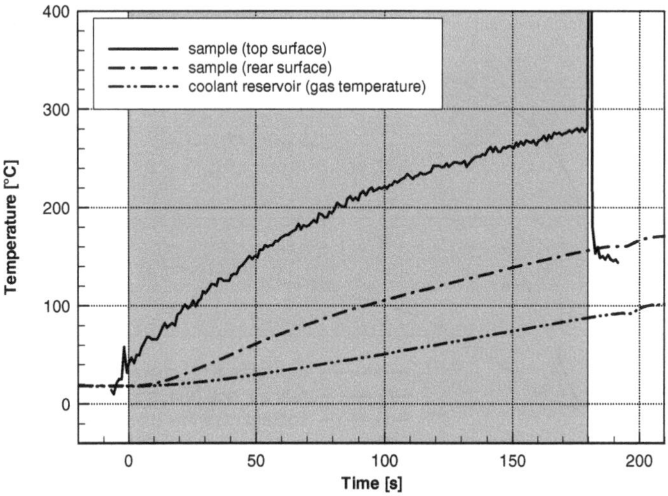

Fig. 23. Temperature measured on sample S1 and in the reservoir (cooled by 0.4 g/s N_2, $\alpha = 20°$)

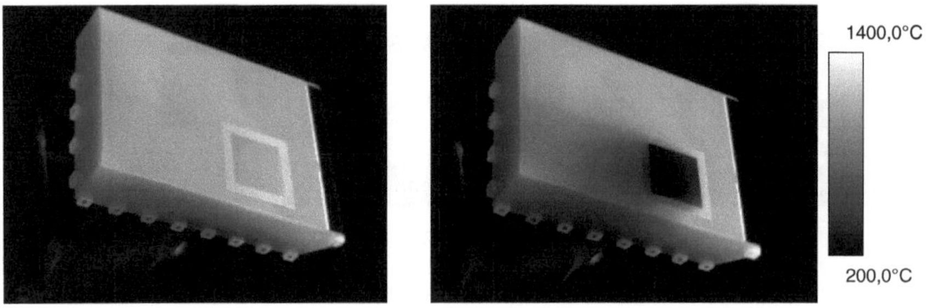

(a) *non-cooled* (b) *cooled by 0.8 g/s N_2*

Fig. 24. Infrared images at the end of a test with and without coolant flow (sample S2, $\alpha = 30°$, flow from right to left)

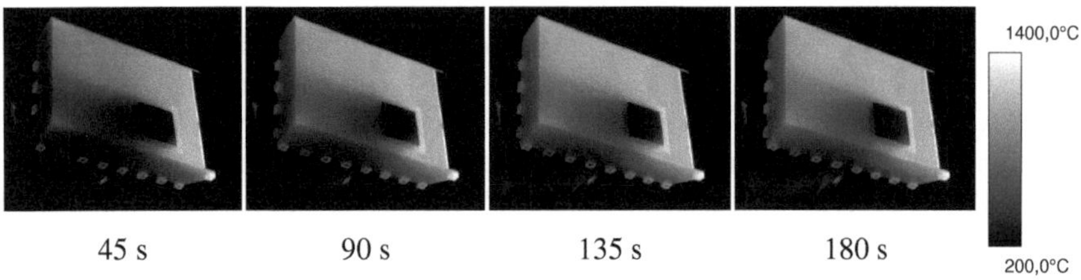

Fig. 25. Sequence of infrared images (Sample S2, cooled by 0.8 g/s N_2, $\alpha = 30°$, flow from right to left)

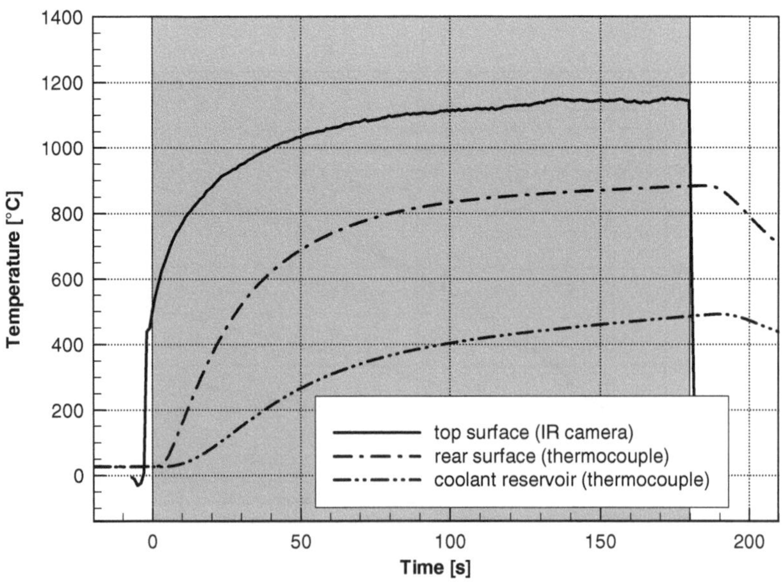

Fig. 26. Temperature measured on sample S2 and in the reservoir during the non-cooled reference test at $\alpha = 30°$

A temperature of 1145°C is reached in the centre of the sample at the end of the non-cooled reference test as shown in Fig. 26. Again, this value is slightly below the corresponding value on the reference side where a value of 1170°C was measured (see Fig. 19). The temperature on the rear side increases to 880°C. Again, there is considerable heating of the reservoir due to hot gas penetration which causes a temperature of close to 500°C at the end of the test.

When cooling the sample with 0.8 g/s nitrogen the surface temperature in the centre of the sample reaches 250°C after 180 seconds which is a reduction by 900 K compared to the non-cooled test (see Fig. 27). At the same time the bottom temperature is 130°C. The gas in the reservoir gradually increases by about the amount than at the lower angle of attack which confirms the assumption that the heat is mainly transferred through the side walls of the reservoir.

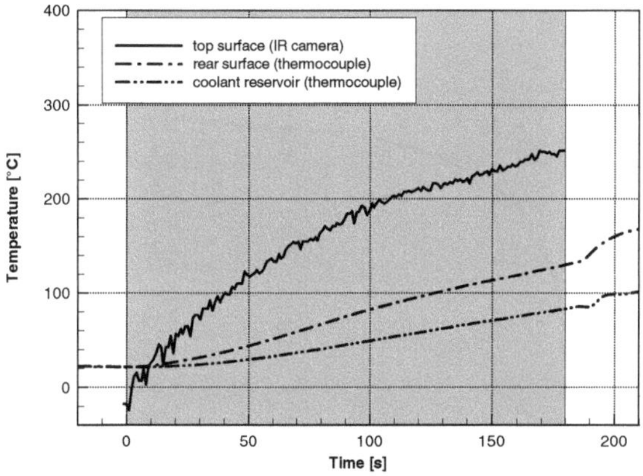

Fig. 27. Temperature measured on the sample and in the reservoir (Sample S2, cooled by 0.8 g/s N_2, $\alpha = 30°$)

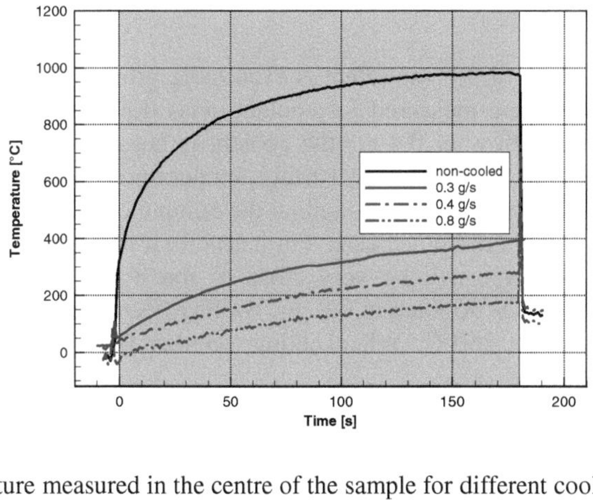

Fig. 28. Temperature measured in the centre of the sample for different coolant mass flow rates (sample S1 and S3, N_2, $\alpha = 20°$)

For the two heat loads different mass flow rates were applied using nitrogen as coolant on"standard-C/C" samples. For $\alpha = 20°$ mass flow rates of 0.3 g/s, 0.4 g/s and 0.8 g/s were tested. The corresponding temperatures that were measured by infrared camera in the centre of the samples are plotted in Fig. 28 and compared to the non-cooled reference test.

Even for the lowest mass flow rate of 0.3 g/s the temperature at the end of the test is remarkably reduced by about 590 K. The final temperature reached is 390°C which is below the material's high temperature limit of operation. By increasing the mass flow rate to 0.4 g/s the final temperature is further reduced by 110°C to 280°C. In order to reduce the temperature by the same amount again, the mass flow rate has to be doubled to 0.8 g/s. The values show that the cooling potential nonlinearly decreases with increasing mass flow rate. While 0.4 g/s of nitrogen enable a temperature reduction of 700 K, another 0.4 g/s added on top have a potential for 110 K only.

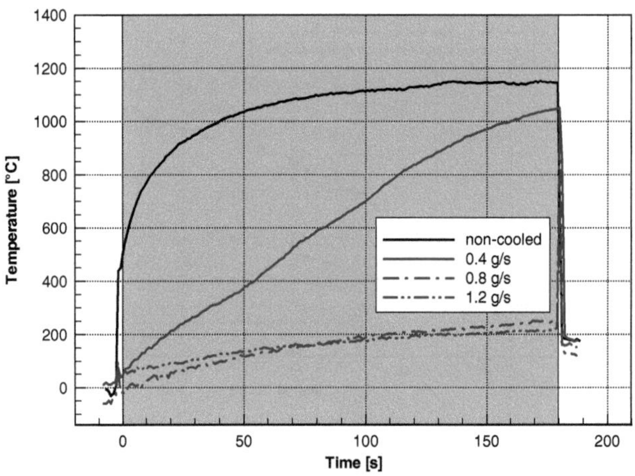

Fig. 29. Temperature measured in the centre of the sample for different coolant mass flow rates (sample S2, N_2, $\alpha = 30°$)

At an angle of attack of 30° the effect is even more pronounced. Figure 29 shows the temperatures that were measured for coolant mass flow rates of 0.4 g/s, 0.8 g/s, and 1.2 g/s. At a mass flow of 0.4 g/s the coolant is not able to reduce the surface temperature permanently. Although the surface temperature is below the temperature that was measured without cooling throughout the complete test, the final temperature is only 100 K below the non-cooled case. Furthermore, it is still increasing indicating that the steady state level will be very close to the non-cooled case. When the coolant's mass flow rate is doubled to 0.8 g/s, cooling becomes effective by reducing the final temperature to 250°C. When adding another 0.4 g/s the temperature is reduced by 33 K only.

The results obtained for different mass flow rates indicate that up to a certain amount transpiration cooling is ineffective, like the 0.4 g/s case in Fig. 29. At high mass flow rates cooling is again ineffective, because a particular amount of coolant is wasted, since it hardly contributes to a reduction of surface temperature. There must be an optimal regime in between with a considerable reduction of surface temperature at moderate mass flow rates. Of course, this optimum depends on the heat load and the porous material.

For $\alpha = 20°$ and "standard-C/C" samples it was tried to assess the optimal mass flow rates for the three available coolants, nitrogen, argon, and helium. During the corresponding tests the mass flow rate was increased gradually and the temperature response of the sample was observed. The optimal mass flow rate was found when the trend of the sample's temperature turned from increasing to decreasing. By that, the optimal mass flow rates were identified to 0.4 g/s for nitrogen, 0.2 g/s for helium, and 0.5 g/s for argon.

Afterwards, regular tests at constant coolant flow were performed for the optimal mass flow rates with an increased testing time of 240 seconds. The corresponding temperature measurements are compared in Fig. 30.

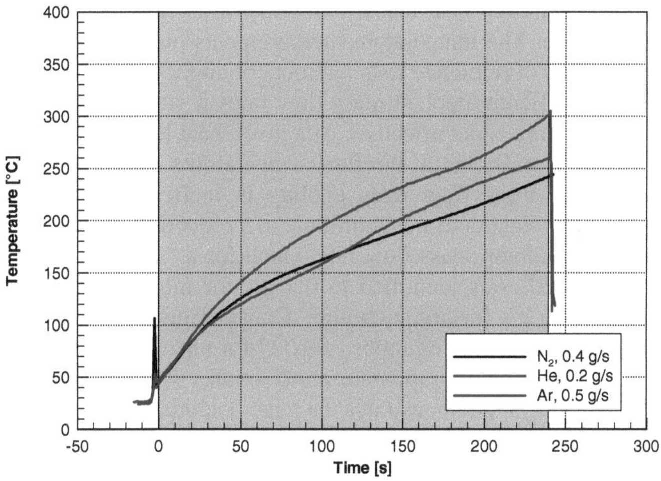

Fig. 30. Temperature measured in the centre of the sample for different coolants at optimal mass flow rates (sample S3, $\alpha = 20°$)

Since the optimal mass flow rates were found by an identical procedure, the temperature histories are similar, but there are still some differences. Although argon was supplied at higher mass flow rate compared to nitrogen and helium, it still provides the highest temperatures on the sample. Nitrogen is performing slightly better than helium at optimal mass flow rate, but helium has the advantage of the lowest mass flow rate.

3.5 Summary and Conclusions

Several test campaigns were performed in the arc heated facilities L2K and L3K in order to qualify the concepts for transpiration cooling that were described in section 6. In two preparatory campaigns useful results were obtained that helped to optimize the experimental setup and the operation of auxiliary systems as well.

For the final test campaign in the L3K facility a flat plate model was used which allowed for integration of square porous samples on one side. The remaining part of the surface was made of C/C-SiC, a well qualified reusable thermal protection material which provided the opportunity to compare the results of transpiration cooling with a qualified conventional thermal protection technology. The model was instrumented for temperature measurements. In addition to thermocouple measurements in the interior the surface was observed by infrared cameras and pyrometers. The porous sample was integrated into a coolant reservoir, which could continuously be supplied by coolant gas from outside the test chamber. A mass flow controller was integrated into the supply line that allowed for constant feeding at specified mass flow rate.

During the test campaign three different porous materials and three different coolants were applied. In addition, the coolant's mass flow rate was varied systematically for one sample material. Tests at two angles of attack, i.e. 20° and 30°, allowed to generate different heat loads. Individual tests were performed at constant coolant mass flow rate. The temperature measurements on the cooled samples were compared to the results of reference tests without cooling.

From the tests at different coolant mass flow rates it was found that for very low rates transpiration cooling is not effective. At a given heat load the increase of surface temperature is delayed only, but at the final temperatures are very close to the non-cooled case. At high mass flow rates cooling is ineffective as well, because a particular amount of coolant does not contribute to a reduction of surface temperature. In between there is optimum mass flow rate which for a "standard-C/C" sample at a model inclination of 20° was identified to 0.4 g/s for nitrogen cooling, 0.2 g/s for helium cooling and 0.5 g/s for argon cooling. Considering the sample's surface area these values correspond to blowing ratios of 0.027 for nitrogen, 0.014 for helium, and 0.034 for argon. These blowing ratios are much smaller smaller than those obtained for film cooling which again demonstrates the huge potential of transpiration cooling.

If only the coolant's mass is taken into account, helium performs best among the three coolants. In case of real applications other parameter might become decisive again, due to limitations to storage volume or reservoir pressures. If e.g. reservoir pressure is limiting variable, the optimum mass flow rates must be weighed by the coolant's molecular mass, yielding 3.96 mole/(m^2s) for nitrogen, 13.9 mole/(m^2s) for helium, 3.47 mole/(m^2s) for argon, and reversing the performance order.

References

1. Esser, B., Gülhan, A., Schäfer, R.: Experimental Investigation of Thermal Fluid/Structure Interaction in High Enthalpy Flow. In: 5th Europ. Symp. on Aerothermodynamics for Space Vehicles, Cologne, ESA SP-563, pp. 275–280 (2004)
2. Gnemmi, P., Schäfer, H.J.: Grundlegende Untersuchungen zur Seitenstrahlsteuerung: Numerische Simulation der Wechselwirkung eines Kaltgasstrahls mit der Außenströmung. ISL, Saint Louis (2004)
3. Gülhan, A., Esser, B.: A Study on Heat Flux Measurements in High Enthalpy Flows. In: 35th AIAA Thermophysics Conference, June 11-14, Anaheim, CA, AIAA (2001)

4. Gülhan, A., Esser, B.: Arc-Heated Facilities as a Tool to Study Aerothermodynamic Problems of Reentry Vehicles. In: Lu, F.K., Marren, D.E. (eds.) Advanced Hypersonic Test Facilities. Progress in Astronautics and Aeronautics, AIAA, vol. 198, pp. 375–403 (2002)
5. Hald, H.: Faserkeramiken für heiße Strukturen von Wiedereintrittsfahrzeugen – Simulation, Test und Vergleich mit experimentellen Flugdaten. Ph.D. thesis, Univ. Stuttgart (2001)
6. Holden, M.S.: A study of flow separation in regions of shock wave – boundary layer interactions in hypersonic flow. AIAA Paper 78-1169 (1978)
7. Jung, K.: Mehrreihige Filmkühlung an gekrümmten Oberflächen. Dissertation, Univ. Darmstadt (2001)
8. Kohli, A., Thole, K.A.: Entrance Effects on Diffused Film-Cooling Holes. ASME, New York (1998)
9. Kovar, A., Schülein, E.: Comparison of experimental and numerical investigations on side jets in a supersonic cross flow. Royal Aeronautical Society, London (2004)
10. Krenkel, W.: Applications of Fibre Reinforced C/C-SiC Ceramics. Ceramic Forum International 80(8), E31–E38 (2003)
11. Lezuo, M.K.: Wärmetransport in H2-transpirativ gekühlten Brennkammerkomponenten, Ph.D. thesis, RWTH Aachen (1998)
12. Meinert, J.: Haftreibung und Wärmeübergang in einer turbulenten Grenzschicht bei Fremdgastranspiration. VDI, Düsseldorf (2000)
13. Radespiel, R., Longo, J.M.A., Brück, S.: Efficient numerical simulation of complex 3D flows with large contrast. In: 77th AGARD Fluid Dynamics Panel, Meeting and Symposium on Progress and Changes in CFD Methods and Algorithms, Sevilla, AGARD-CP-578, pp. 1–33 (1995)
14. Serbest, E.: Untersuchungen zur Anwendung der Effusionskühlung bei Raketenbrennkammern., Ph. D. Thesis, RWTH Aachen (2002)
15. Wittig, S., Thole, K.A.: Transonic Film-cooling Investigations: Effects of Hole Shapes and Orientations, ASME, New York, 35/96-222 (1996)

Experimental Study of Active Cooling in 8 Laminar Hypersonic Flows

K.A. Heufer and H. Olivier

Stoßwellenlabor
RWTH Aachen
Templergraben 55, 52056 Aachen, Germany
heufer@swl.rwth-aachen.de

Summary

In this study fundamental investigations concerning film cooling on an inclined flat plate and on a stagnation point region in laminar flow are performed. The influence of the relevant parameters, i.e. the Reynolds number of the free stream, the blowing ratio and the geometry of the blowing opening, is investigated by experiments. The results show the weak influence of the cooling gas injection on the outer flow. But the experimental data achieved in a shock tunnel demonstrate that already low blowing rates lead to a significant reduction of the thermal loads. Overall the results exhibit that the film cooling technique is an effective method for cooling bodies in laminar hypersonic flows.

1 Introduction

Most of today's thermal protection systems (TPS) for hypersonic flow conditions can be described as passive, i.e. they can not be adjusted during flight and they do not exert direct influence to the flow around the body to reduce thermal loads. One famous example of this kind of TPS is the heat shield of the Space Shuttle. Here materials with very low thermal conductivities are used to protect the main structure from the heat loads during reentry [3]. For the different regions along the vehicle and therewith for the different heat loads the TPS is adapted so far, that diverse materials and wall thicknesses are used to reduce the weight of the TPS as much as possible. An alternative to such a system can be the film cooling technique. Thereby a cooling gas is introduced into the flow near the wall, which should establish a cooling film over the structure surface to reduce the penetrating heat flux. This technique is already state of art for cooling turbine blades. Here mostly air is used as cooling gas which is led through discrete holes on the surface of the blades. A comprehensive overview of the influence of various parameters is given in Refs. 6 and 14. In newer concepts, the active cooling technique is used for cooling rocket combustions chambers, where hydrogen is led through porous walls into the combustion chamber. Additionally to the gas film, a cooling effect already establishes in the porous wall [8, 15]. For hypersonic flow conditions already in the 60's first papers with experimental and theoretical results concerning active cooling methods have been published [2, 5, 7]. But these studies do not give any results for the case of high enthalpy hypersonic

A. Gülhan (Ed.): RESPACE – Key Tech. for Reusable Space Sys., NNFM 98, pp. 132–150, 2008.
springerlink.com

flows, where chemical reactions take place. However, all of these show that film cooling technique is an effective method to lower the thermal loads of a structure.

In addition, the past shows that the reliability of a thermal protection system is of great importance for flight safety. This can be increased by a suitable active film cooling system, since in case of a local failure of the system this can be compensated by sufficient cooling capacity upstream of the damaged part of the thermal protection system. Thus the superior goal of this study is to examining the possibility of applying an active cooling system for bodies in laminar hypersonic flows, thereby aiming at a low structural weight realized by a high effectiveness of the system and additionally increasing its reliability.

In this study, first fundamental investigations of injecting a cooling gas in hypersonic flows are done with a simple inclined flat plate with transverse slots. The advantage of this is given by the two-dimensionality of the flow under ideal conditions. The flow parameters for such a configuration without cooling were examined in numerous analytical, numerical and experimental studies in the past, so that for this case the flow conditions on the inclined flat plate are well known [1, 4]. The experiments have been performed in the shock tunnel TH2 of the Shock Wave Laboratory, RWTH Aachen University. Ideally the cooling gas should establish a film over the surface which separates the hot flow from the wall. For film cooling the cooling results from two effects. On the one hand the boundary layer thickness is increased by the added mass of the cooling gas. This leads to smaller velocity and temperature gradients at the wall. On the other hand the heat flux to the wall can be reduced by the cooling gas parameters like temperature, heat capacity and heat transfer coefficient.

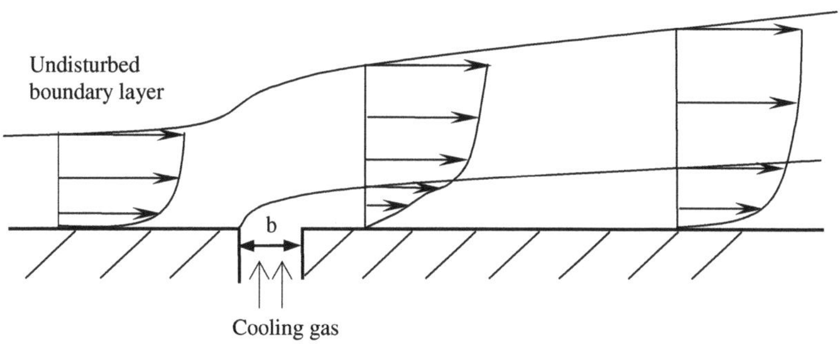

Fig. 1. Principle of film cooling

The different conditions in the cooling gas and the boundary layer flow of the outer gas lead to gradients between these two gases. With an increasing distance from the blowing opening these gradients reduce thus increasing the temperature and the velocity gradients at the wall (Fig. 1). This effect and an expected mixing of the two flows reduces the cooling effectiveness with increasing distance from the slot. In contrast to this an increasing cooling effect should be received by an increasing cooling mass flow. The effectiveness of a cooling method can be defined in different ways. In literature concerning cooling of turbine blades the following definition of the cooling effectiveness for a stationary flow is common

$$\eta = \frac{T_r - T_{aw,c}}{T_r - T_{0,cg}}. \tag{1}$$

Here $T_{aw,c}$ indicates the adiabatic wall temperature in the case of cooling, T_r the recovery temperature of the free stream and $T_{0,cg}$ the total temperature of the cooling gas. In the experiment, the recovery temperature can be determined by the free stream conditions. The total temperature of the cooling gas is equal to its reservoir temperature. The evaluation of the adiabatic wall temperature in the case of cooling is more difficult due to the short test times of the employed shock tunnel in which the adiabatic temperature is not reached during the experiment. In this case the heat flux to the wall can be measured and has to be transferred into an equivalent temperature. The transformation is done by a similar approach given by Holden [13]. In general the heat flux to the surface can be expressed as follows

$$\dot{q} = \lambda \cdot (T_{aw} - T_w). \tag{2}$$

The changes of the wall temperatures during the experiments are small due to the short testing times of the shock tunnel, so that the wall temperature can be assumed to be constant at room temperature. Applying equation 2 for both cases with (c) and without (nc) cooling leads to an expression for the adiabatic wall temperature in the case of cooling

$$T_{aw,c} = T_{aw,nc} - \frac{\dot{q}_{nc}}{\lambda_{nc}} + \frac{\dot{q}_c}{\lambda_c}. \tag{3}$$

For the case of no cooling the adiabatic wall temperature is equal to the recovery temperature of the free stream. Moreover, for the present experimental conditions the reservoir temperature of the cooling gas is also equal to the room temperature thus leading to a simple formula of the cooling effectiveness by inserting equation 3 into 1

$$\eta = 1 - \frac{\dot{q}_c}{\dot{q}_{nc}} \cdot \frac{\lambda_{nc}}{\lambda_c}. \tag{4}$$

At last an assumption for the ratio of the heat transfer coefficients has to be made. In general the heat transfer coefficient mainly depends on Nusselt, Prandtl and Reynolds number. In case that the flow is not significantly disturbed by the injection of the cooling gas, the flows with and without cooling are of same kind, i.e. the flow over the surface is still a laminar boundary flow. Furthermore, for the considered experiments the cooling gas properties are very similar to those of the free stream so that in conclusion the ratio of the heat transfer coefficients can be assumed to be approximately one leading to

$$\eta = 1 - \frac{\dot{q}_c}{\dot{q}_{nc}}. \tag{5}$$

In [11] it has been proven by numerical simulations for isothermal and adiabatic wall conditions that the definition of the cooling effectiveness given by equation 5 suitable for the flow conditions in impulse facilities yields the same results as for adiabatic wall conditions (Eq. 1) representing flight conditions.

Additionally to the cooling effectiveness more parameters have to be defined for the discussion of the experimental results. One of the main parameters of the cooling process is the blowing ratio F which is defined as the ratio of the specific mass flow of the cooling gas and the free stream

$$F = \frac{\rho_c u_c}{\rho_\infty u_\infty}. \tag{6}$$

The thermal loads to the surface can be expressed by the Stanton number

$$St = \frac{\dot{q}}{\rho_\infty u_\infty (h_0 - h_w)}. \tag{7}$$

Pressure measurements are presented in the usual form of the pressure coefficient

$$c_p = \frac{p - p_\infty}{\frac{\rho_\infty}{2} u_\infty^2}. \tag{8}$$

2 Wind Tunnel Model and Test Facility

For most of the experiments an inclined flat plate with a sharp leading edge is used. Due to this setup the test surface is inclined to the free stream by 30° (Fig. 2). The test surface is equipped with up to 30 coaxial thermocouples and 20 pressure gauges to determine the heat flux and pressure distribution, respectively. In order to allow additionally infrared measurements black plastic materials (PVC) with a high emission factor was chosen as test surface material.

Beside the one slot configuration with an orthogonal slot in respect of the surface (Fig. 1) different parameter configurations have been investigated whereby the blowing ratio, the blowing angle, the slot number, the slot width and the Reynolds number of the free stream have been varied. The test plate has a width of 350 mm and a length in stream-wise direction of 250 mm. Furthermore, for some experiments the slot blowing opening has been replaced by a field of blowing holes and the sharp leading edge by a rounded one to study the cooling at a stagnation region of blunt bodies. The experiments concerning the cooling of the stagnation region where done with a cylindrical nose part (Fig. 3). The cylindrical part of the nose has a diameter of 60 mm. Each configuration is described in detail in the corresponding chapter.

For the experiments the shock tunnel TH2 of the Shock Wave Laboratory is employed. The shock tunnel can be operated with a helium or detonation driver [9, 16]. For both operation modes flow conditions have been calibrated. The typical range of flow conditions is listed in Table 1.

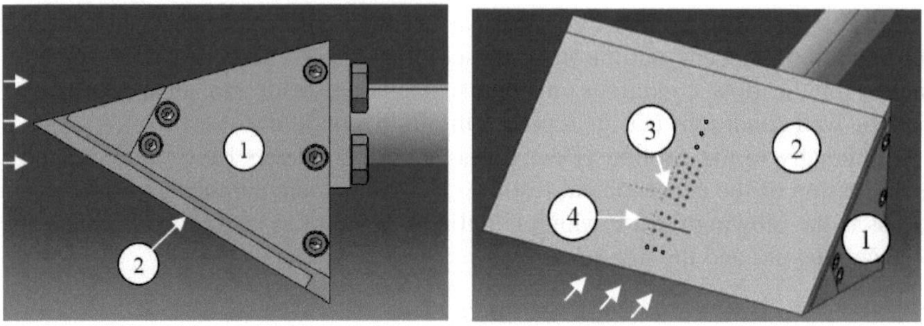

Fig. 2. Model with sharp leading edge 1) support structure, 2) test plate, 3) probe positions, 4) blowing slot

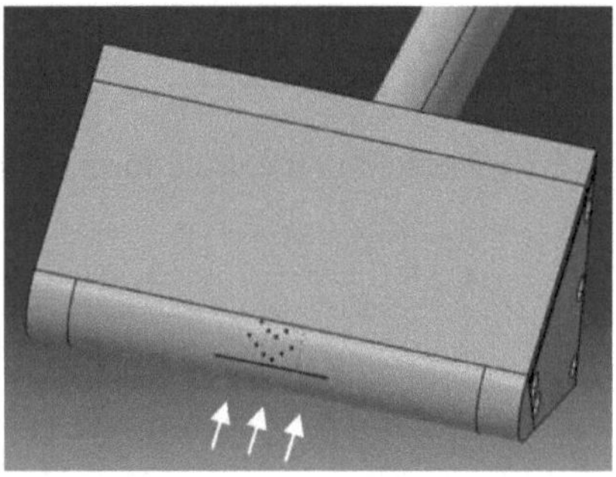

Fig. 3. Model with cylindrical front part

Table 1. Conditions of the shock tunnel TH2

Stagnation values		Free stream conditions	
p_0 [bar]	50 - 560	Ma_∞	6.5 – 12.0
T_0 [K]	1100 - 7400	Re_∞ [10^6 m^{-1}]	0.8 – 16.5
h_0 [MJ/kg]	1.1 – 14.2	u_∞ [m/s]	1400 - 4850
		T_∞ [K]	80 - 1300
		p_∞ [bar]	0.004 – 0.3
Measuring time [ms]	1 - 8	Test gas	Air

For the experiments described in this paper two similar conditions are employed, condition I and Ih, listed in Table 2. The difference between the two conditions is caused by a certain amount of air added to the helium driver gas for condition Ih. For this first set of experiments a low enthalpy condition has been chosen in order to avoid the influence of high temperature real gas effects which allows to separately study the film cooling behaviour in nonreacting and reacting flows.

To determine the blowing ratio F the density and flow velocity at the boundary layer edge at the slot position are required. In order to achieve these values the flow conditions behind the front shock of the model are calculated assuming ideal gas

Table 2. Free stream and post shock values a) condition I b) condition Ih

	Condition I	Post shock		Condition Ih	Post shock
T_0 [K]	1390	1390	T_0 [K]	1110	1110
u_∞ [m/s]	1660	1234	u_∞ [m/s]	1470	1150
T_∞ [K]	102	564	T_∞ [K]	80	486
ρ_∞ [kg/m^3]	0.0179	0.0908	ρ_∞ [kg/m^3]	0.0193	0.0985
Re_∞ [10^6 m^{-1}]	4.2	3.9	Re_∞ [10^6 m^{-1}]	5.3	4.3
Ma_∞	8.2	2.7	Ma_∞	8.3	2.6

behaviour, which is justified by the relatively low total temperatures of the test flow (Table 2).

3 Experimental Results

3.1 Experiments without Cooling

The experiments without cooling yield the reference heat flux distribution that is needed for the determination of the cooling effectiveness described in equation 5. Furthermore, these tests give the possibility to compare the heat fluxes and Stanton numbers respectively with empirical values found in literature [1] to validate the experimental setup. In figure 4a the Stanton number distribution found from the experiments and the results of an empirical formula are plotted versus the distance x from the leading edge of the flat plate. The experimental values found from the thermocouple measurements range between ±10% of the average value and show a good agreement to the empirical formula as well as the Stanton number distribution deduced from the infrared measurements.

Figure 4b shows the Stanton number distribution for the cylindrical nose model versus the peripheral angle of the cylindrical part. Just as good as for the flat plate a good agreement is obtained between the literature values and the experiment.

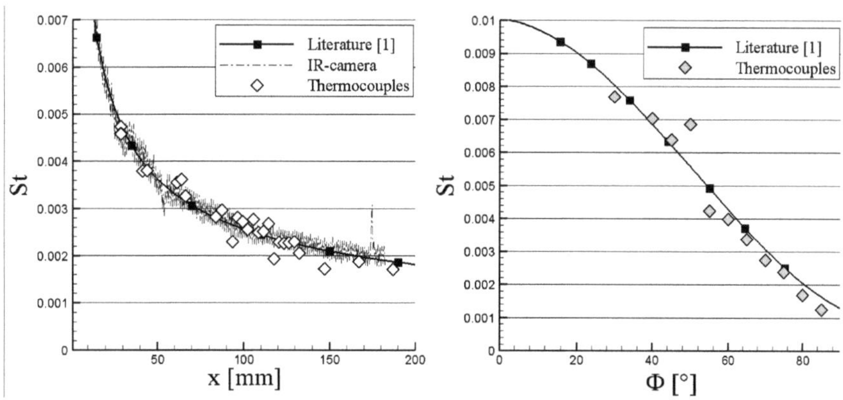

Fig. 4. Reference heat flux distribution for the case without cooling left: sharp leading edge, right: cylindrical nose

3.2 Influence of the Cooling Gas Injection on the Flow Field

For first investigations a slot blowing opening with a width of $b = 0.5$ mm and a spanwise length of $t = 60$ mm arranged normal to the surface has been used. The distance of the slot from the leading edge amounts to $x_1 = 55$ mm (Fig. 5).

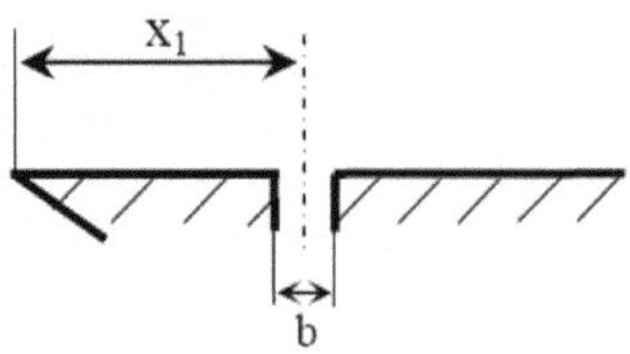

Fig. 5. Single slot arranged normal to the surface, $x_1 = 55$ mm, b = 0.5 mm

Regarding ideal film cooling the disturbance of the flow by the injection of the cooling gas should be small to ensure an optimum cooling effectiveness. A strong blowing ratio would induce a strong bow shock in front of the blowing opening [10], thus leading to a higher static pressure and higher thermal loads behind this shock. Figure 6 shows an example for the ratio of the pressure coefficients with and without cooling versus the distance *x* from the leading edge. For the presented blowing ratios as well as for all other blowing ratios and slot configurations considered the pressure distribution does not change significantly. This leads to the conclusion that no strong shocks are induced by the cooling gas injection.

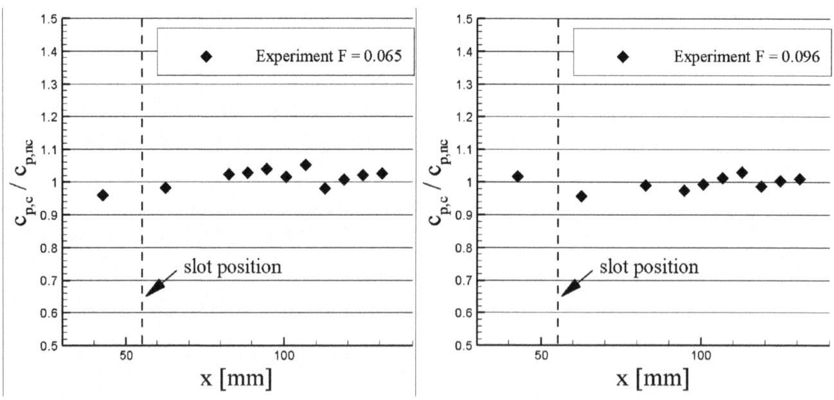

Fig. 6. Influence of the cooling gas injection on the pressure distribution for two different blowing ratios

Schlieren pictures taken during the experiment confirm this conclusion. Corresponding to figure 1 at the slot position first compression then expansion waves should appear due to the deflection of the flow. Assuming weak compression and expansion waves respectively the angle of these waves in respect of the surface can be estimated by the Mach angle

$$sin\,\sigma = \frac{1}{Ma}. \tag{9}$$

Calculating the Mach number behind the front shock of the model by the ideal oblique shock relations results in a Mach angle of σ = 22.7°. Figure 7 shows a Schlieren photograph for the slot configuration shown in Fig. 5. The expected waves are visible at the slot position and the angle of these waves measured from the Schlieren picture approximately amounts to 23°. This validates the assumption of weak pressure waves initiated by the cooling gas injection.

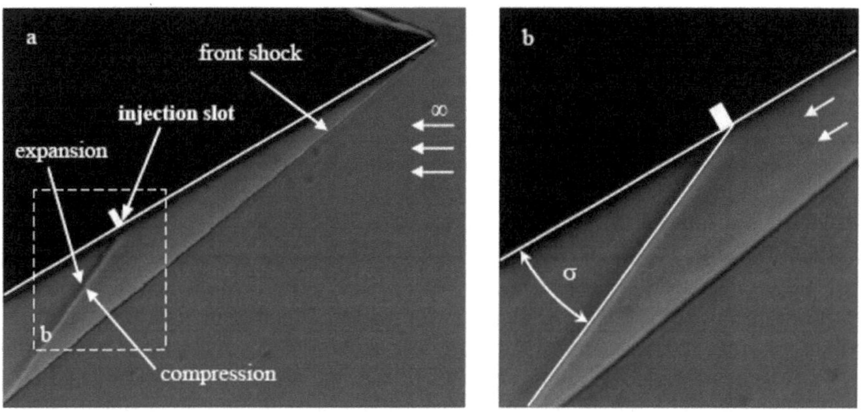

Fig. 7. Influence of the cooling gas injection visualized in a Schlieren picture, F = 0.065

The main aim of the experiments concerning the film cooling technique is to investigate the reduction of the thermal loads due to the cooling gas injection. Figure 8 shows a typical result of the film cooling experiments. On the left side the measured heat flux is shown derived from thermocouple and infrared measurements in comparison to the heat flux distribution for the case without cooling versus the distance from the leading edge x of the flat plate. Immediatly upstream of the blowing opening the heat flux decreases and reaches its minimum at the slot position. With an increasing distance from the blowing opening the heat flux increases again until it approaches the heat flux distribution for the case without cooling. Applying equation 5 to this result leads to the graph presented on the right side of Fig. 8 where the cooling effectiveness is plotted versus the distance x. A zero cooling effectiveness represents no cooling and an effectiveness of one the maximum achievable cooling, i.e. a reduction of the wall heat flux to zero. The increasing cooling effectiveness immediately in front of the blowing opening and the corresponding decrease of the wall heat flux respectively can be explained by the detachment of the main flow due to a separation induced by the cooling gas injection. Downstream of the blowing opening as expected the gradients between the main flow and the cooling film reduce with increasing distance from the slot so that the cooling effectiveness decreases.

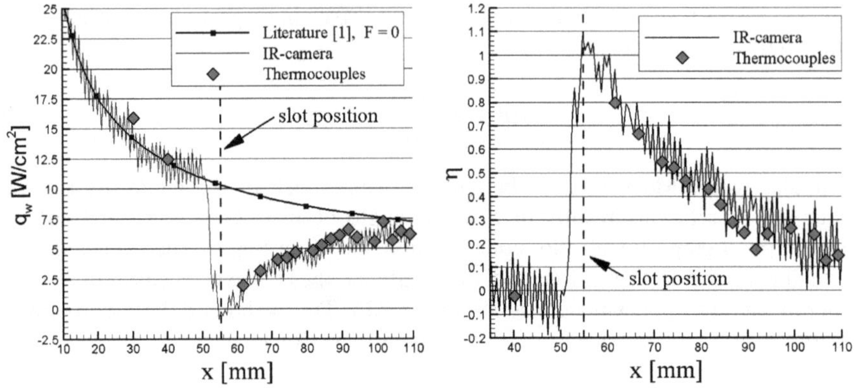

Fig. 8. Reduction of the thermal loads due to cooling gas injection, F = 0.096

3.3 Influence of the Blowing Ratio F

The influence of the blowing ratio F has been investigated with the configuration described in Figure 5. To interpret the experimental results correctly, the behaviour of the flow before the useful measuring time starts, has to be investigated. Due to the short test times in the shock tunnel TH2 it is not possible to switch on the cooling mass flow during the experiment. Moreover, the experimental setup requires an evacuated test section so that already prior to the experiment the blowing of the cooling gas into the vacuum tank of the test section starts. This leads to a certain pressure in the so called blowing chamber of the injection system which is installed directly beneath the blowing opening. Thereby the pressure in this chamber depends on the cooling mass flow rate and the geometry of the blowing slot. When the experiment starts, the static pressure of the flow increases leading to a pressure increase in the blowing chamber. For very low cooling mass fluxes the pressure rise in the blowing chamber is not fast enough so that no constant cooling mass flux during the test time can be established. The corresponding blowing ratio is defined as the minimum blowing ratio F_{min}. In the opposite, also a maximum blowing ratio F_{max} can be defined. Strong disturbances due to high blowing ratios may cause transition from a laminar to a turbulent boundary layer. As seen in Fig. 9 for $F > F_{max}$ the wall heat flux downstream of the injection position increases rapidly and the cooling effectiveness becomes negative, i.e. the wall is stronger heated up than for a laminar flow. For these blowing ratios larger than F_{max} the infrared camera measures a hot area behind the slot (Fig. 9) which indicates a boundary layer transition.

Within these limits of the blowing ratio ($F_{min} \approx 0.03$ and $F_{max} \approx 0.15$) the experiments show a significant cooling effect. Figure 10 shows the cooling effectiveness plotted as a function of the blowing ratio F for different positions *x*. As expected the cooling effectiveness increases with an increasing blowing ratio and it decreases with an increasing distance from the injection slot. At low blowing ratios a

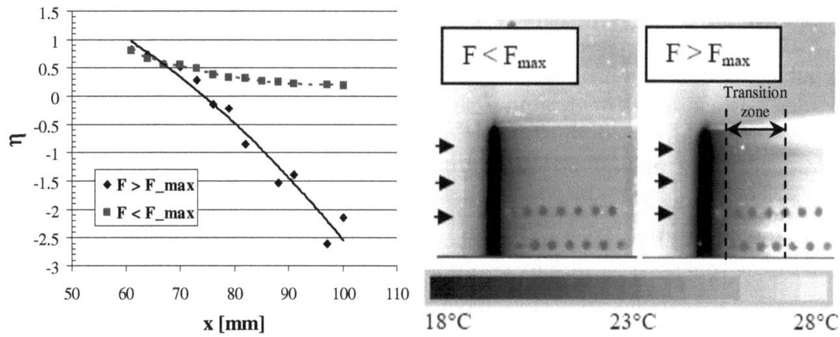

Fig. 9. Boundary layer transition due to high blowing ratios, left: results from thermocouple measurement, right: infrared thermography of surface temperature

strong increase of the cooling effectiveness with an increasing blowing ratio can be observed while at higher blowing ratios only a slight increase appears until boundary layer transition occurs. This can be explained by the generation of some turbulence caused by the injection of the cooling gas. This reduces the cooling effectiveness due to the mixing of the hot outer flow and the cooling gas. At low blowing ratios this turbulence is of small influence, but the influence increases at higher blowing ratios. Near the blowing opening the induced disturbances are stronger and become weaker with increasing distance from the slot. If a critical disturbance level at high blowing ratios is reached, the disturbances do not decay and a transition to a fully turbulent flow occurs.

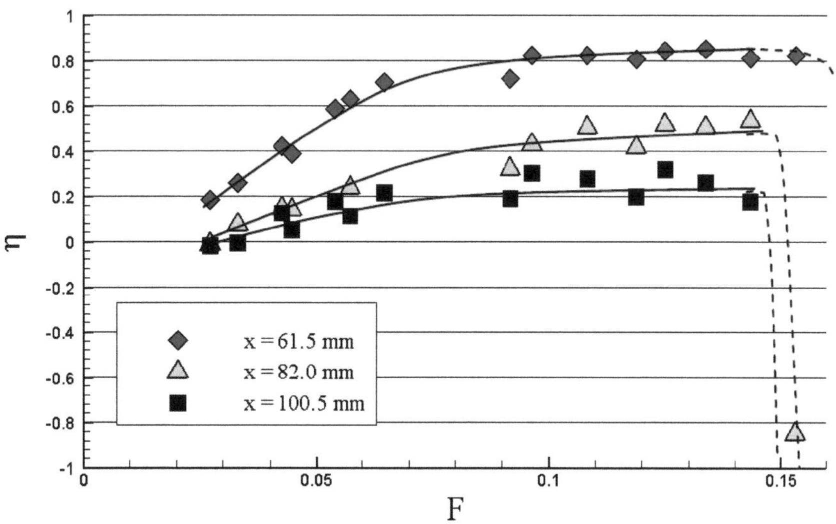

Fig. 10. Influence of blowing ratio on cooling effectiveness

3.4 Experiments with Multislot Arrangement

The double slot configuration was designed to investigate the possibility of increasing the cooled surface area by arranging several slots one after each other. At each slot the basic behaviour of a single slot configuration is expected. If the blowing openings are arranged close enough to each other it should be possible to reestablish the cooling process with low blowing ratios. Depending on the distance Δx between the slots the blowing ratio of each opening can be set to establish a desired minimum cooling effectiveness η_{min} for any cooling length.

For these experiments two slots have been arranged one after each other with a distance of 20 mm (Fig. 11).

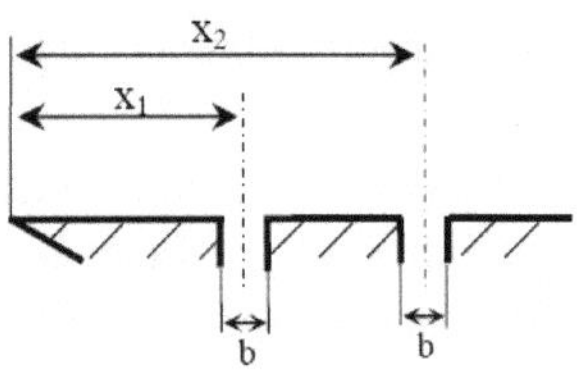

Fig. 11. Double slot model, b = 0.5 mm, x_1 = 55 mm, x_2 = 75 mm

The experimental results reveal that the cooling effectiveness for a double slot configuration with small blowing ratios can be approximated quite well from the single slot experiments. This is done by superimposing the cooling effectiveness of the single slot configuration for the two slot positions of the double slot configuration. Thereby, Reynolds number effects have been neglected because of the small distance between the two slots. Figure 12 shows exemplarily the comparison between the deduced distribution and the measured one for different blowing ratios. For the blowing ratios investigated the experiments fit well to the deduced distribution.

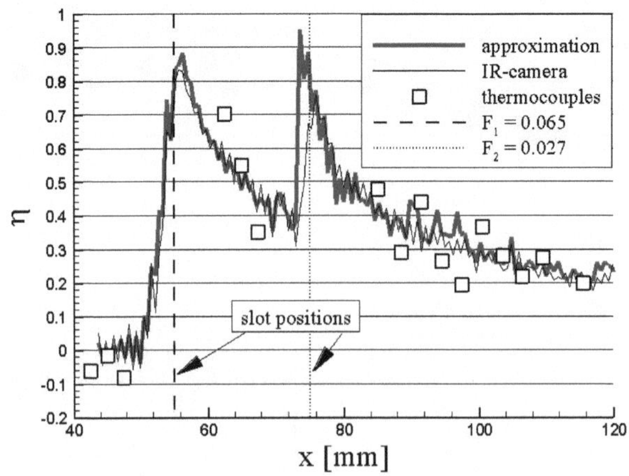

Fig. 12. Comparison between measured and deduced cooling effectiveness for double slot configuration

3.5 Influence of the Slot Width

To investigate the influence of the slot width the single slot configuration with one slot arranged normal to the surface has been used with two different slot widths of $b_1 = 0.25$ mm and $b_2 = 0.5$ mm (Fig. 13). The experimental results show that to achieve the same cooling effectiveness the double blowing ratio for the slot width b_1 in comparison to the slot width b_2 is needed (Fig. 14). This leads to the conclusion that within the investigated range the slot width has no influence on the cooling effectiveness when the absolute cooling mass flow is constant. That means the ratio of the slot widths can be used as a scaling factor for the blowing ratio. Furthermore, it can be concluded that within the investigated range for constant cooling mass flow ($bF = const.$) the cooling effectiveness is independent from the injection gas velocity.

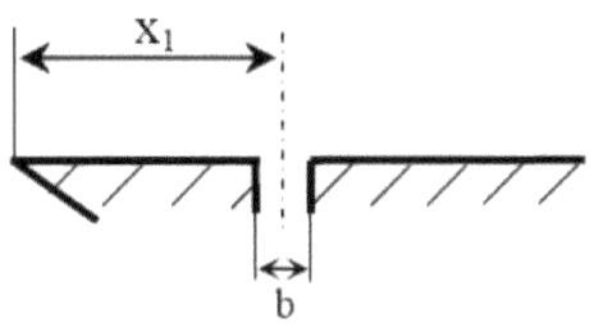

Fig. 13. Single slot, $x_1 = 55$ mm, $b_1 = 0.25$ mm, $b_2 = 0.5$ mm

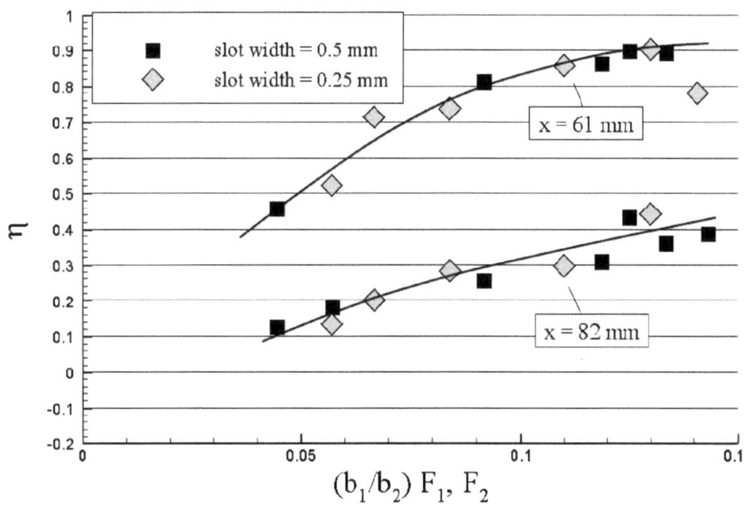

Fig. 14. Influence of the slot width and scaling

3.6 Influence of the Blowing Angle

For the oblique slot configuration the angle of the blowing opening in respect of the surface was chosen to 45° (Fig. 15). The results of the slot width variation already showed that the cooling effectiveness is independent from the blowing velocity

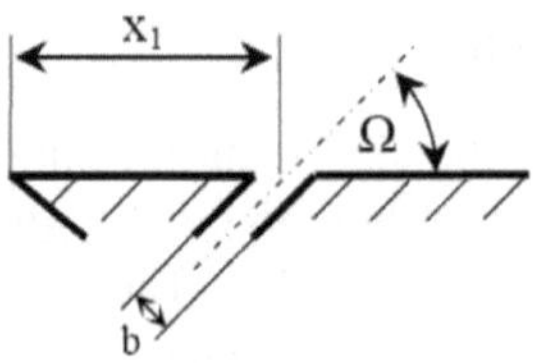

Fig. 15. Oblique slot, $x_1 = 55$ mm, b = 0.5 mm, Ω = 45°

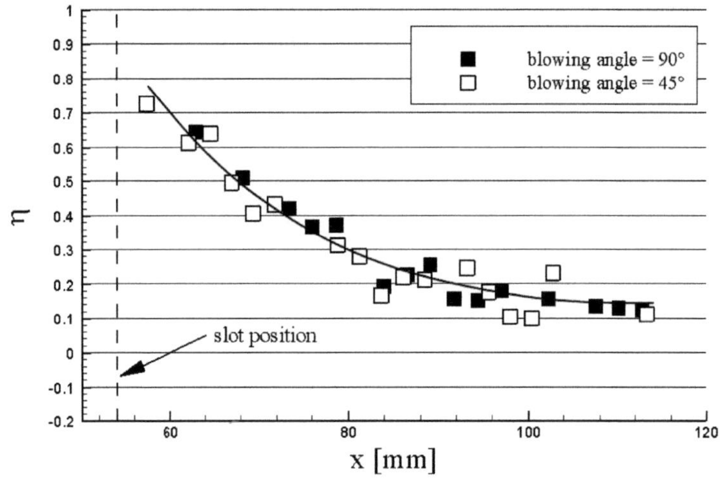

Fig. 16. Comparison between normal and oblique slot blowing, F = 0.065

normal to the wall so that for the case of an oblique slot no influence of the blowing angle is expected. Figure 16 shows this exemplarily for a blowing ratio of $F = 0.065$ for the two investigated blowing angles.

3.7 Influence of the Reynolds Number

The Reynolds number effect has been investigated by varying the distance from the leading edge to the blowing opening. For these experiments the normal slot configuration has been chosen for two distances from the leading edge x_1 and x_2,

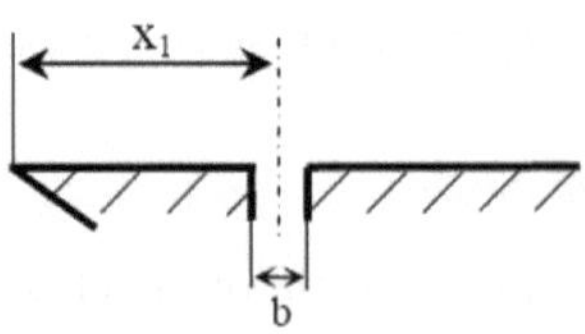

Fig. 17. Single slot, $x_1 = 55$ mm, $x_2 = 125$ mm

respectively (Fig. 17). These reference lengths lead to Reynolds numbers of $2.4 \cdot 10^5$ and $5.4 \cdot 10^5$ respectively.

Thereby the boundary layer thickness of the undisturbed boundary layer at the slot position x_1 increases to about 50% and the reference wall heat flux decreases about 33% compared to the slot position x_1. In contrast to this for the investigated Reynolds number range the experimental results presented in Figure 18 do not show a significant influence of the Reynolds number.

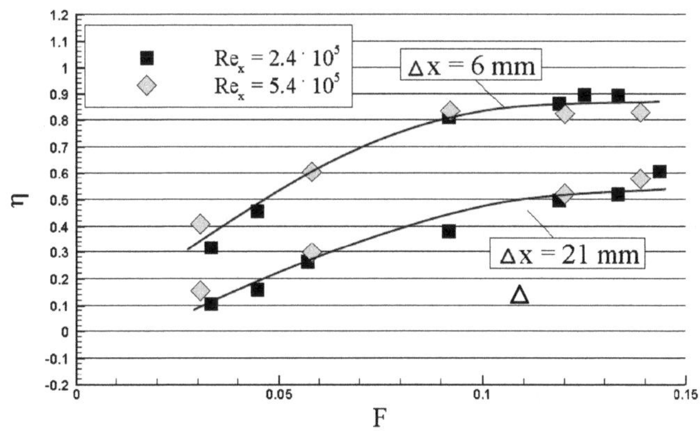

Fig. 18. Influence of the Reynolds number for two distances from the blowing opening Δx

3.8 Experiments with a Field of Blowing Holes

The goal of the experiments with a field of blowing holes was to investigate the possibility of replacing the slot blowing opening by an alternative configuration, because a slot blowing opening is likely difficult to integrate into the structure of a real vehicle.

To achieve a uniform cooling film two rows of blowing holes with a distance in spanwise direction between each other smaller than their diameter has been used (Fig. 19). As desired the infrared pictures show a uniform temperature distribution behind the blowing field, which on the infrared picture is not resolved in detail because of the ejecting cooling gas (Fig. 20). The comparison of the cooling effectiveness between the slot blowing opening and the field of blowing holes shows that in spite of the expected behaviour the blowing field is not significant less effective than the slot opening for low blowing ratios (Fig. 21). This result and those achieved for different slot widths and blowing angles suggest that in a laminar flow the cooling effectiveness is independent from the blowing geometry.

But the blowing field has a huge effect on the maximum blowing ratio. For the slot blowing opening this maximum blowing ratio is larger than $F_{max} > 0.14$, for the field of blowing holes already at a blowing ratio of $F = 0.077$ a boundary layer

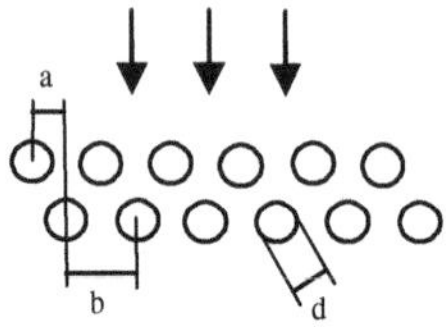

Fig. 19. Field of blowing holes, a = 0.4 mm, b = 2·a, d = 0.5 mm

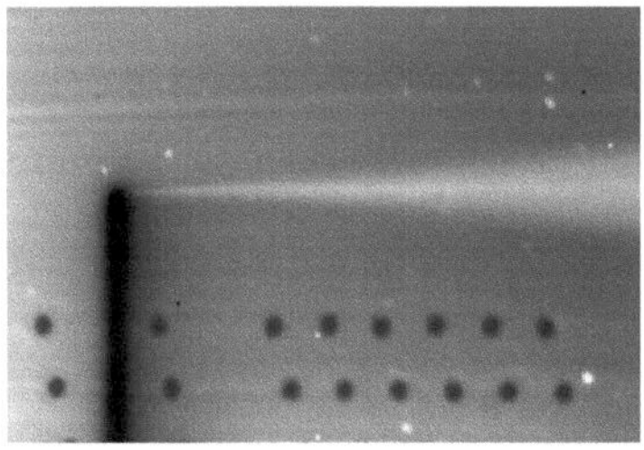

Fig. 20. Infrared picture of blowing field experiment

transition occurs. The lower maximum blowing ratio for the blowing field configuration can be explained by the additional disturbances due to the vortices generated at each blowing hole which is not the case for the slot blowing opening. This results in a lower maximum achievable cooling effectiveness at a constant distance from the blowing opening.

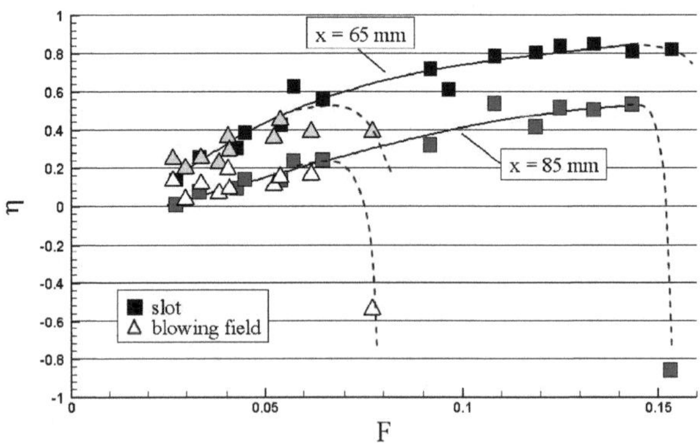

Fig. 21. Comparison between slot blowing opening and field of blowing holes

3.9 Cooling at the Stagnation Region

In respect of the flat plate the flow around a cylindrical body is characterized by a negative pressure gradient and a stagnation region. The flow along the stagnation stream line decelerates to zero and accordingly the pressure rises. Outward of the stagnation point the flow accelerates again and the pressure drops around the cylinder. Due to the low velocity in the stagnation region the impulse of the main stream that is responsible for the deflection of the cooling gas flow is also low. Thus for high blowing rates and an injection normal to the wall the cooling gas penetrates deeply into the main flow and therefore does not establish a cooling film (Fig. 22). For lower blowing rates the velocity of the cooling gas is very low and the cooling gas is heated up before it can build up a cooling layer a certain distance downstream of the injection position. Thus for a stagnation point cooling a porous material, a field of blowing holes or a more tangential blowing angle in respect of the surface should be applied to ensure a cooling effect over a larger area. To overcome these problems, in the presented work the injection slot is arranged at an angle of $\Phi = 30°$ outside of the stagnation point. In this region the main flow has been accelerated to a certain amount so that here a positive effect of film cooling is expected again. For this configuration the blowing ratio is calculated with the free stream conditions in front of the bow shock.

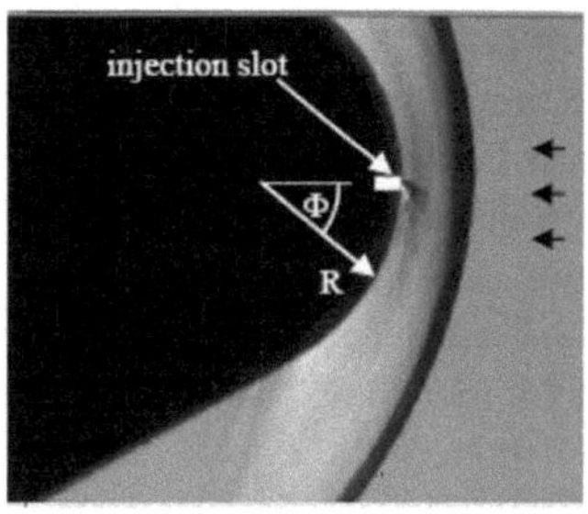

Fig. 22. Schlieren picture of injection at the stagnation region, slot inclined to the free stream 1°, R = 30 mm

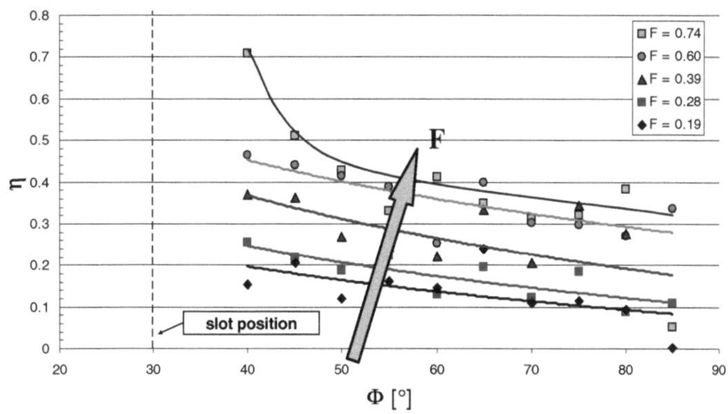

Fig. 23. Cooling effectiveness around the cylindrical nose for different blowing ratios

Figure 23 shows the experimentally determined cooling effectiveness around the cylindrical nose of the model. In principle these results give the same behaviour as the results achieved from the flat plate experiments. With an increasing distance from the blowing opening the cooling effectiveness decreases and it increases with an increasing blowing ratio. Near the blowing opening a high cooling effectiveness and a strong increase of the effectiveness with an increasing blowing ratio is observed. More downstream the dependency of the cooling effectiveness on the cooling gas mass flow becomes weaker at high blowing ratios. In contrast to the experiments with the inclined flat plate no boundary layer transition has been observed for the highest investigated blowing ratio although it is about an order of magnitude higher. This can be due to an effect of the stabilizing negative pressure gradient around the cylinder. Furthermore, for the same blowing ratio for the cylindrical nose the blowing impulse is less due to the higher pressure level and therefore higher density and lower velocity respectively of the cooling gas.

3.10 Three-Dimensional Effects Due to Finite Injection Slot Length

Comparing the results of two-dimensional CFD simulations [12] with experiments presented above a difference in the achievable cooling effectiveness is observed. While the qualitative behaviour of the cooling effectiveness versus the different influencing parameters is the same for all slot lengths being investigated, the absolute measured cooling effectiveness is somewhat lower (Fig. 24). Experiments with larger slot lengths in spanwise direction reveal that the flow with a slot length of $t = 60$ mm cannot be regarded as fully two dimensional, i.e. vorticy structures at the edges of the slot (see IR-pictures) seem to have an influence on the midspan measuring section. Experiments with slot lengths larger than $t = 90$ mm show a good agreement to the numerical simulation while reducing the slot length leads to a lower cooling effectiveness.

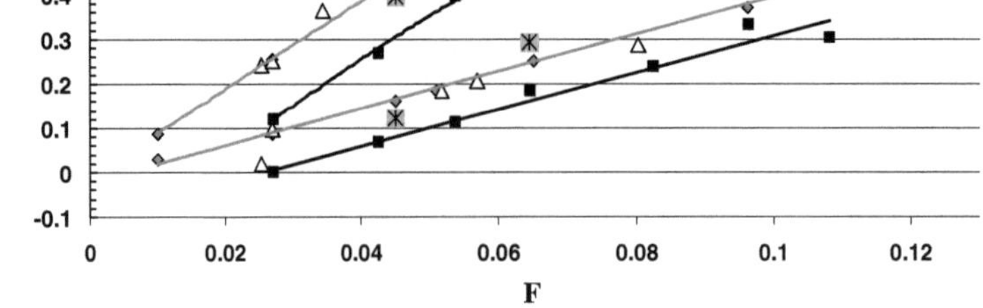

Fig. 24. Influence of the finite slot length on cooling effectiveness

4 Summary

In the presented project the basic behaviour of film cooling for hypersonic laminar flow conditions has been investigated. Most of the experiments have been performed with an inclined flat plate with a slot blowing opening to achieve a quasi two-dimensional flow. The experiments with a single slot show a significant cooling effect at relatively low cooling mass flows. As expected the cooling effect can be increased by increasing the cooling mass flow and the effect decreases with an increasing distance from the blowing opening. Thereby it is important to notice that at high blowing ratios a boundary layer transition may occur. This would lead to higher heat fluxes as for the case without cooling and thus has to be avoided. This can be achieved by a larger number of blowing openings with lower blowing ratios whereby the width of the slots is of no influence on the cooling effectiveness as long as the cooling mass flow is constant. The alternative to slot blowing opening, a field of blowing holes, shows no significant lower cooling effect but a lower maximum blowing ratio. This configuration likely is more important for real applications because of the better integration into a structure. Finally, experiments concerning cooling a stagnation region have been presented and for a region outside of the stagnating region mainly the same qualitative behaviour of the cooling effectiveness as for the flat plate case has been shown.

Recent results published in [12] show that for the inclined flat plate configuration all of the presented data can be correlated quite well by a suitable correlation parameter which takes into account the main physical effects governing the problem of film cooling in hypersonic laminar flow.

Additionally to the reduction of the thermal loads, the overall mass for an active cooling system is of interest for a real application. As shown before, the distribution of the cooling effectiveness for multiple slots arranged one after each other can be approximated from the results of a one slot configuration. With this result a first approximation for the weight of an active cooling system can be made. As reference the conditions for the critical reentry phase between 80 and 55 km of the space shuttle is chosen. The time of operation of the active cooling system during the critical reentry phase is estimated by 13 minutes and the blowing ratio based on the presented experimental results is chosen to yield a reduction of the thermal loads by 50%. Based on this data, compared to the existing thermal protection system a total weight reduction of about 20% of the TPS can be estimated. This impressively demonstrates the high potential of this technology.

References

1. Anderson, J.D.: Hypersonic and high temperature gas dynamics. AIAA 2000 (2000) ISBN 1-56347-459-X
2. Cresci, J.R., Libby, P.A.: The downstream influence of mass transfer at the nose of a slender cone. J. of the Aerospace Sciences 29, 815–826 (1962)
3. Curry, D.M.: Space shuttle orbiter thermal protection system design and flight experience. NASA-TM-104773 (1993)
4. van Driest, E.R.: Investigation of a laminar boundary layer in compressible fluid using the Crocco method. NACA TN2597 (1952)
5. Fox, H., Libby, P.A.: Helium injection into the boundary layer at an axisymmetric stagnation point. J. of the Aerospace Sciences 29, 921–934 (1962)
6. Goldstein, R.J.: Film cooling. In: Irvine, T.F., Hartnett, J.P. (eds.) Advances in Heat Transfer, vol. 7, pp. 321–379. Academic Press, New York (1971)

7. Gollnick Jr., A.F.: Thermal effects on a transpiration cooled hemisphere. J. of the Aerospace Sciences 29, 583–595 (1962)
8. Greuel, D., Serberst, E., Haidn, O.J.: Einsatz von porösen Faserkeramiken für Effusionskühlung im Düsenbereich von Raketenbrennkammern. Deutscher Luft- und Raumfahrtkongress, DGLR-JT2001-118 (2001)
9. Grönig, H., Olivier, H.: Experimental hypersonic flow research in Europe. JSME Int. J., Series B 41(2), 397–407 (1998)
10. Gruber, M.R., et al.: Mixing and penetration studies of sonic jets in a Mach 2 freestream. J. of Propulsion and Power 11(2), 315–323 (1995)
11. Heufer, K.A., Olivier, H.: Film Cooling of an Inclined Flat Plate in Hypersonic Flow. In: AIAA Paper 2006-8067, 14th AIAA/AHI Space Planes Conference and Hypersonic Systems Conference, Canberra, Australia (2006)
12. Heufer, K.A., Olivier, H.: Experimental and numerical investigation of film cooling in hypersonic flows. In: 26th International Symposium on Shock Waves, Paper-ID 3091, Göttingen, Germany (2007)
13. Holden, M.S.: Experimental studies of shock-wave/wall-jet interaction in hypersonic flow. NASA-CR-195844 (1994)
14. Jung, K.H.: Mehrreihige Filmkühlung an gekrümmten Oberflächen. Ph.D. Thesis, University Darmstadt (2001)
15. Lezuo, M.K.: Wärmestransport in H2-transpirativ gekühlter Brennkammer. Ph.D. Thesis, RWTH Aachen (1998)
16. Olivier, H., et al.: Detonation-driven shock tubes and tunnels. In: Lu, F., Marren, D. (eds.) Advanced Hypersonic Test Facilities, Progress in Astronautics and Aeronautics, vol. 198, pp. 135–203. AIAA Inc. (2002)

Numerical Investigations of Film Cooling

J. Linn and M.J. Kloker

Institut für Aero- und Gasdynamik
Universität Stuttgart
Pfaffenwaldring 21, 70550 Stuttgart, Germany
Jens.Linn@iag.uni-stuttgart.de

Summary

Effusion cooling by discrete slits and holes in various laminar zero-pressure-gradient super- and hypersonic boundary layers is investigated using direct numerical simulation. A comparison with experimental data for a Mach 2.67 boundary layer with a cool wall and a spanwise slit shows good agreement. For an adiabatic Mach 6 boundary layer it was found that slits are better than holes due to the lower blowing velocity. Slit blowing causes a destabilisation of 2nd mode disturbances, and a complete stabilisation of 1st modes despite the generated maxima of the spanwise vorticity inside the boundary layer. Hole blowing gives rise to counter-rotating streamwise vortices, with a noticeable laminar-flow destabilisation only for large spanwise hole spacings. For a radiation-adiabatic wall at flight conditions the principal behavior is similar but part of the cooling efficiency is lost because of the decreased radiation of heat.

1 Introduction

For aerospace or hypersonic cruise vehicle the state of the boundary layer is of great importance because for turbulent boundary layers, the thermal loads and skin friction are higher than in laminar boundary layers. Therefore, knowledge of cooling features and laminar-turbulent transition is necessary for the design of the thermal protection system (TPS). Different strategies are used to reduce the thermal loads of hypervelocity vehicles, e.g. radiation, ablation, transpiration or effusion cooling.

Direct numerical simulations (DNS) are carried out to investigate the effect of effusion cooling by blowing through spanwise slits and discrete holes in a laminar flat-plate boundary layer at various Mach numbers. The numerical method and boundary conditions are described in section 2. A comparison with experimental data of project RESPACE-A7 for a Mach 2.67 boundary layer with a cool wall and a spanwise slit is presented in section 3.1. In section 3.2, a comparison of various effusion-cooling configurations (slits and holes) and laminar instability investigations are made for an adiabatic Mach 6 boundary layer. Results for effusion cooling at flight conditions are shown in section 3.4, and section 4 summarizes the results.

A. Gülhan (Ed.): RESPACE – Key Tech. for Reusable Space Sys., NNFM 98, pp. 151–169, 2008.
springerlink.com

2 Numerical Method

2.1 Governing Equations

The numerical method is based on the complete 3-d unsteady compressible Navier-Stokes equations, continuity equation and the energy equation. These equations can be written in dimensionless form as:

$$\frac{\partial \rho}{\partial t} + \nabla \cdot (\rho \vec{u}) = 0 \tag{1}$$

$$\frac{\partial (\rho \vec{u})}{\partial t} + \nabla \cdot (\rho \vec{u} \vec{u}) + \nabla p = \frac{1}{Re} \nabla \vec{\sigma} \tag{2}$$

$$\frac{\partial (\rho e)}{\partial t} + \nabla \cdot (p + \rho e) \vec{u} = \frac{1}{(\kappa - 1) Re\, Pr\, Ma^2} \nabla \cdot (\vartheta \nabla T) + \frac{1}{Re} \nabla \cdot (\vec{\sigma} \vec{u}) \tag{3}$$

where

$$\vec{\sigma} = \mu \left[\left(\nabla \vec{u} + \nabla \vec{u}^T \right) - \frac{2}{3} (\nabla \cdot \vec{u}) I \right] \tag{4}$$

is the viscous stress and

$$e = c_v \cdot T + \frac{1}{2} \left(u^2 + v^2 + w^2 \right) \tag{5}$$

is the internal energy per mass unit. The air is considered as a non-reacting calorically perfect gas [3, 10]

$$p = \frac{1}{\kappa Ma^2} \cdot \rho T \tag{6}$$

with a constant Prandtl number ($Pr = 0.71$) and specific heat ratio of $\kappa = c_p / c_v = 1.4$. The viscosity is calculated using Sutherland's law [11].

All length scales are dimensionless with respect to a reference length

$$L^* = \frac{\nu_\infty^* \cdot Re}{u_\infty^*}. \tag{7}$$

Reference values for velocity, density temperature, viscosity and conductivity are their freestream values at the inflow (indicated by subscript ∞). The pressure is normalised by $\rho_\infty^* u_\infty^{*\,2}$, where the superscript * denotes dimensional quantities. With these definitions, the global and running-length Reynolds numbers are respectively defined as

$$Re = \frac{u_\infty^* \cdot L^*}{\nu_\infty^*} = 10^5 \tag{8}$$

and

$$Re_x = \frac{u_\infty^* \cdot x^*}{\nu_\infty^*} = x \cdot 10^5 . \tag{9}$$

2.2 Spatial and Time Discretisation

The Navier-Stokes equations are solved in a rectangular integration domain (Fig. 1) on the flat plate, well below the shock wave induced by the leading edge. In streamwise (*x*-) and wall-normal (*y*-) direction, the discretisation is realized by splitted compact finite differences of 6th order [6]. In the spanwise (*z*-) direction, the flow is assumed to be periodic, thus a Fourier spectral representation is employed. The time integration is done with a classical 4th-order Runge-Kutta method. A detailed description of the discretisation and algorithm is reported in [1].

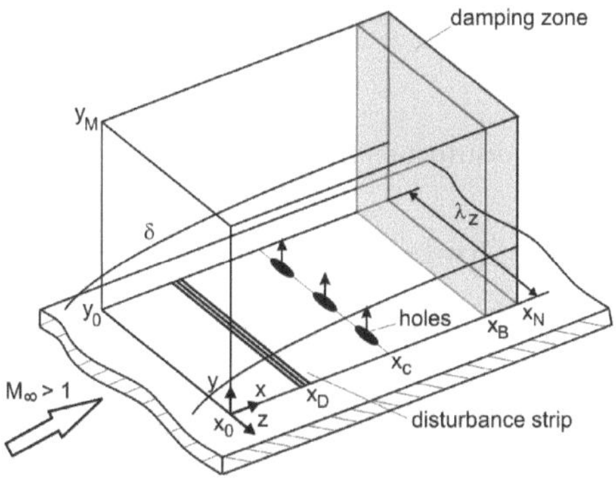

Fig. 1. Integration domain

2.3 Initial and Boundary Conditions

The numerical simulation is performed in two steps. First, the steady base flow is calculated by solving the Navier-Stokes equations using a pseudo time stepping for integrating the time-dependent equations to a steady state. For real unsteady simulation this base flow is used as initial state ($t = 0$). Disturbance waves are introduced for $t > 0$ by localized periodic blowing and suction in a disturbance strip, and the spatial downstream development of the disturbance waves is calculated from the full equations. We use a disturbance flow formulation, meaning that all flow quantities are splitted in their base-flow and disturbance part ($\phi = \phi_{BF} + \phi'$), to ease the formulation of specific boundary conditions. The full equations are used and a non-linear generated time mean is contained in the disturbance flow ($< \phi' > \neq 0$).

At the inflow boundary ($x = x_0$), profiles from boundary-layer theory are fixed for all variables, and the disturbances are zero. For the base-flow boundary condition at the outflow ($x = x_N$), all equations are solved neglecting the second x-derivative terms and for the disturbance flow, all disturbances are damped to zero in a damping zone shortly upstream the outflow boundary. At the freestream boundary ($y = y_M$) for the base flow, the gradient of the flow variables is set to zero along spatial characteristics [3]. An exponential decay condition is used for the disturbance flow [10].

At the wall, all velocity components are zero, except within the slits, holes, and disturbance strip. The steady blowing of cold air through holes at the wall with a radius r_c (Fig. 2) is modelled by prescribing a wall-normal mass-flux distribution

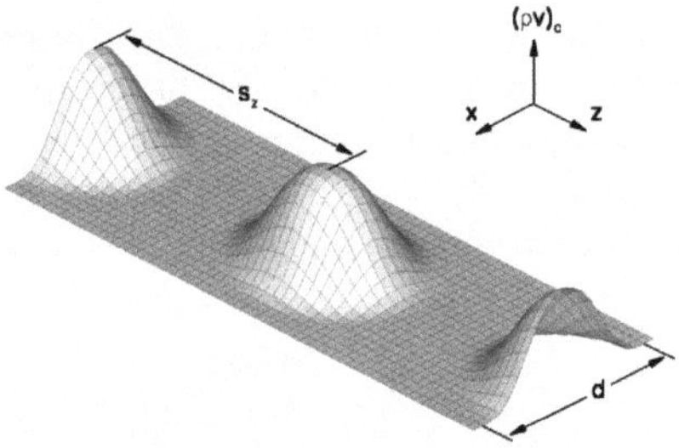

Fig. 2. (ρv)-distribution at the wall for one row of holes

$$(\rho v) = (\rho v)_{c,max} \cdot c(r) \tag{9}$$

where $(\rho v)_{c,max}$ is the maximum blowing ratio. The wall temperature distribution over the blowing is prescribed by

$$T_c = T_w \cdot (1 - c(r)) + T_{c,core} \cdot c(r) \tag{10}$$

where $T_{c,core}$ is the core temperature of the cold air and T_w is the local wall temperature at the edge of the hole. The distribution function $c(r)$ is a polynomial of 5th order [7], which has already been used in [9] for suction and blowing to generate disturbances at the wall. Both the gradient and curvature are zero at $r = 0$ and $r = r_c$

$$c(r) = 1 - 6 \cdot \left(\frac{r}{r_c}\right)^5 + 15 \cdot \left(\frac{r}{r_c}\right)^4 - 10 \cdot \left(\frac{r}{r_c}\right)^3 \tag{11}$$

$$r = \sqrt{(x - x_c)^2 + (z - z_c)^2}, \quad 0 \le r \le r_c \tag{12}$$

where x_c and z_c stand for the center coordinates of the hole. Outside the hole, (ρv) is zero at the wall. T_w is outside the blowing region either the local adiabatic wall temperature

$$\left.\frac{\partial (T_{BF} + T')}{\partial y}\right|_w = 0 \tag{13}$$

or has a constant value (isothermal wall; $T_{BF} = const$; $T' = 0$), or is the radiation-adiabatic wall-temperature

$$\vartheta \left. \frac{\partial (T_{BF} + T')}{\partial y} \right|_w = \varepsilon \sigma T_w^4 \quad \text{with} \quad \varepsilon = 0.8 . \tag{14}$$

The pressure gradient in wall-normal direction at the wall, holes, slits and disturbance strip is zero. For steady blowing through a spanwise slit of width $b_c = 2 \cdot r_c$, the distribution function $c(r)$ in Eq. (11) is independent from the z-coordinate, i.e. $z = z_c$. Disturbance waves are introduced within a disturbance strip by time-periodic simultaneous blowing and suction, which is modelled by a distribution of the wall-normal mass flux $(\rho v)'$ over the strip [3].

Grid-refinement studies varying Δx, Δy, Δz and Δt for exemplary cases were performed and convergence was always found. The corresponding results are discussed below.

3 Results

3.1 Comparison with Experiments at Mach 2.67

In this section, we compare our simulation results with the measurements by Heufer and Olivier at the Shock Wave Laboratory Aachen (SWL; RESPACE-A7). They investigated an isothermal laminar boundary layer on a wedge with a deflection angle of 30° and a given post-shock Mach number 2.67, with $M_\infty = 7$ and $T_\infty = 1368$ K. Cold air is blown through one spanwise slit. The given post-shock freestream temperature is $T_{ps}^* = 564$ K (≈ 0.45 T_{rec}^*), the pressure $p_{ps}^* = 0.1489$ bar ($\rightarrow L^* = 24.57$ mm, $Re_{unit}^* = 4.07 \cdot 10^6$ 1/m), and the wall temperature is $T_w^* = 293$ K = *const* (≈ ¼ T_{rec}^*). This wall temperature means that the wall itself is already strongly cooled. Investigations based on the Linear Stability Theory (LST) have shown (see, e.g. [8]), that wall cooling stabilises 1st-mode (vorticity) disturbances and destabilises the 2nd-mode (acoustic) disturbances that do not exist at Mach numbers lower than approximately 3.5. In addition, the basic boundary layer investigated is subcritical ($R_{x,crit} >> R_x(x_N)$) due to the strong wall cooling.

The core temperature of the effusion air is $T_{c,core}^* = 293$ K ($= T_w^*$) and the slit width is $b_c = 0.5$ mm corresponding to $0.57 \cdot \delta_c$, where δ_c is the boundary-layer thickness without blowing at the slit position. The cooling effectiveness is defined by

$$\eta_c = 1 - \frac{\dot{q}_c}{\dot{q}_{ref}} \tag{15}$$

where $\dot{q}_c = \lambda \cdot \partial T / \partial y|_w$ is the heat flux into the wall with effusion cooling and $\dot{q}_{ref}$ without effusion cooling.

The cooling effectiveness behind the slit is shown over the blowing rate in Fig. 3. It increases linearly with the blowing rate in the simulation and is slightly lower than in the experiment. Note that no experimental data of the boundary-layer evolution are available and thus the local thickness parameters and Reynolds numbers may differ. Lower cooling effectiveness was also found by tentative numerical simulations at SWL.

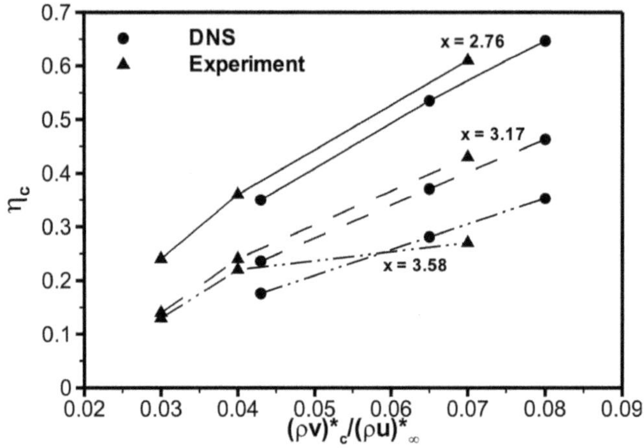

Fig. 3. Cooling effectiveness η_c from simulation and experiment as function of the blowing rate at three downstream positions for an effusion-cooled boundary layer at Mach 2.67 (line with dots – simulation; lines with triangle - experiment)

A longitudinal cut of the temperature field with streamlines is shown in Fig. 4 for the blowing rate $(\rho v)^*_c/(\rho v)^*_\infty = 0.065$. In front of the slit is a reversed-flow region with a clockwise rotating vortex with its center marked by the dot. For both blowing rates, no instability regions were found using spatial LST despite a separation region exists in front of the slit. The basic cooling by the cool wall is so strong that it

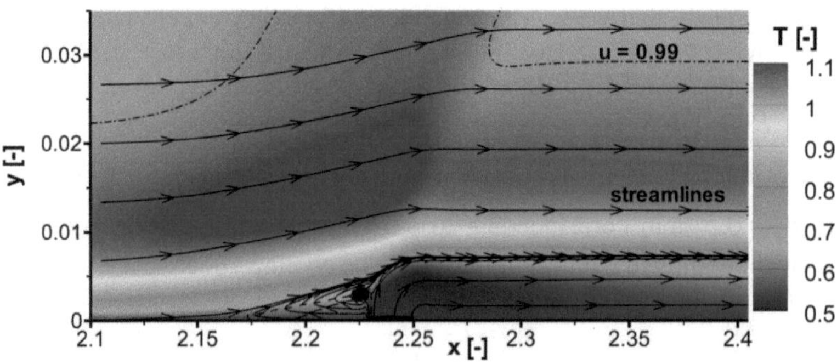

Fig. 4. Visualisation of the temperature field with streamlines in a longitudinal cut for the effusion-cooled boundary layer at Mach 2.67 ($(\rho v)_{c,max} = 0.065$). Isolines of the u-velocity for u = 0 (dashed line) and u = 0.99 (dashed dotted line). $\Delta x = 0.25\ 10^{-2}$ and $\Delta y = 0.6\ 10^{-3}$.

stabilizes even the blowing. Here effusion cooling is applied in a case where the flow is already strongly cooled by a cool wall, thus this case is unrealistic. A simple transfer of the results to cases with significantly different wall temperature gradients is not possible as DNS not presented here have shown.

3.2 Comparison of Effusion-Cooling Configurations at Mach 6

In this section we investigate an adiabatic boundary layer at an edge Mach number 6 in which cold air is blown through spanwise slits and rows of holes. The freestream temperature is $T^*_\infty = 89$ K ($\approx {}^1/_7\ T^*_{rec}$) and the pressure is $p^*_\infty = 0.0038$ bar ($\rightarrow L^* = 36.28$ mm, $Re^*_{unit} = 2.8 \cdot 10^6$ 1/m), matching the flow parameters of experiments in the hypersonic wind tunnel H2K of DLR-Köln [2] (RESPACE-A6). Table 1 summarizes the parameters. Two successive slits were used in case A, piecewise homogeneous blowing (one wide slit) in case B and holes in cases C, D. The integrally injected mass flow and the cooling gas temperature $T_{c,core} = 293$ K ($\approx {}^1/_2\ T^*_{rec}$) are in all cases identical. In case C, the two rows of holes are aligned in contrast to case D where the rows are $s_z/2$ staggered.

Table 1. Parameters of the slit and hole configurations for cases at $M_e = 6$

case	$(\rho v)_{c,max}$	hole diameter or slit width d	streamwise spacing s_x	spanwise spacing s_z	rows z-offset
A	0.0284	$0.055 \approx 0.56\ \delta_c$ = 2 mm	$0.1378 \approx 1.4\ \delta_c$ = 5 mm	–	–
B	0.00984	0.193	–	–	–
C	0.15	0.055	0.1378	0.1378	–
D	0.15	0.055	0.1378	0.1378	$s_z/2$

$x_0 = 0.225$, $x_N = 7.33$, $y_M = 0.54 \approx 4\ \delta_c$ at $x = x_N$, blowing starts at $x_c = 2.205$ ($x_c^* = 80$ mm)

The resulting wall temperature is shown in Fig. 5. Cases A and B show a significant lower T_w than the two other cases. The "homogeneous" blowing model with its low wall-normal velocity (case B) has the lowest T_w. (We remark that, due to the used model, T_w would keep low even if the blowing vanished.) In the „aligned" case C, T_w is only slightly reduced and strongly varies in the z-direction. In the „staggered" case D, T_w is lower than in case C, and does not vary as strong in the z-direction. The temperature and u-distribution in the crosscut is shown in Fig. 6. The aligned rows blow more cold gas from the wall into the boundary layer and show stronger $\partial u/\partial z$-gradients than the staggered rows. The reason why the slits are more efficient is that the blowing surface is much larger than with the holes, translating into

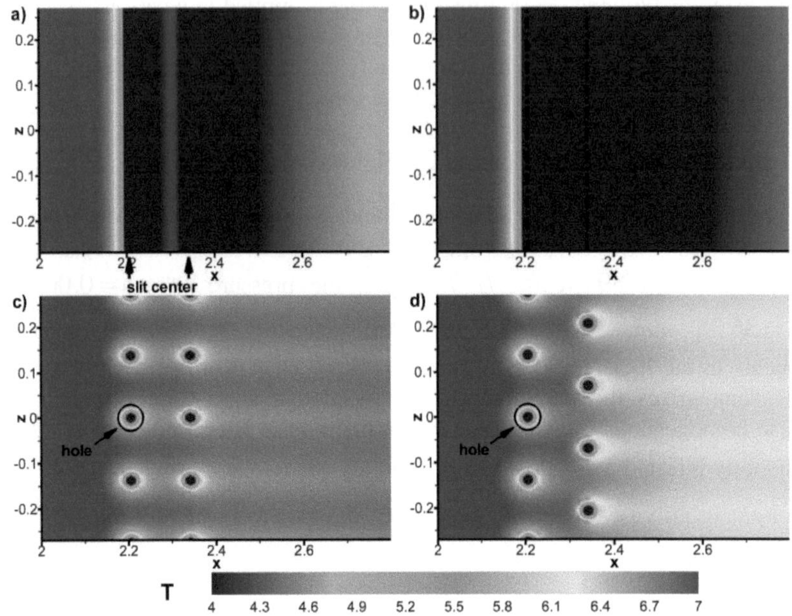

Fig. 5. Wall temperature for steady blowing into an adiabatic flat-plate boundary layer at Mach 6 through A two spanwise slits, B piecewise homogeneous blowing (one wide slit), C two aligned rows of holes, and D two $s_z/2$-staggered rows of holes

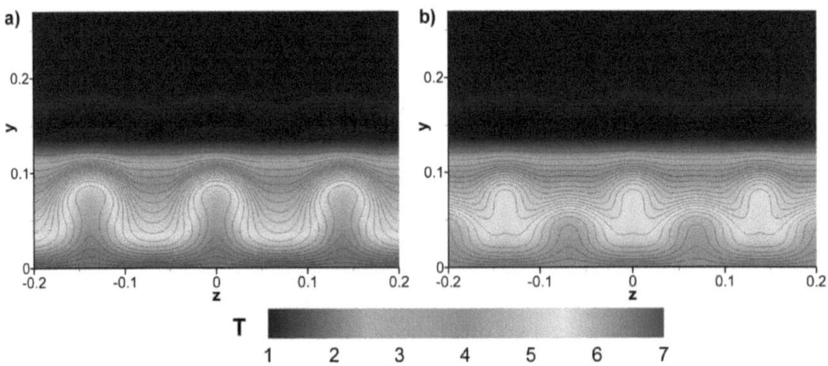

Fig. 6. Visualisation of the temperature field and isolines of the u-velocity in the crosscut at x = 2.5 for the aligned (A) and for the $s_z/2$-staggered rows of holes (B)

a lower wall-normal velocity in the slits. Thereby the cold gas keeps closer to the wall. Decreasing the spanwise and streamwise spacing of the holes increases the cooling effectiveness.

The vortical structures of the hole configurations are visualised via the λ_2-criterion [5] in Figs. 7 and 8. From the holes, counter-rotating vortex pairs (CVPs) emerge which are along the jet trajectory and have such a rotation sense that gas is

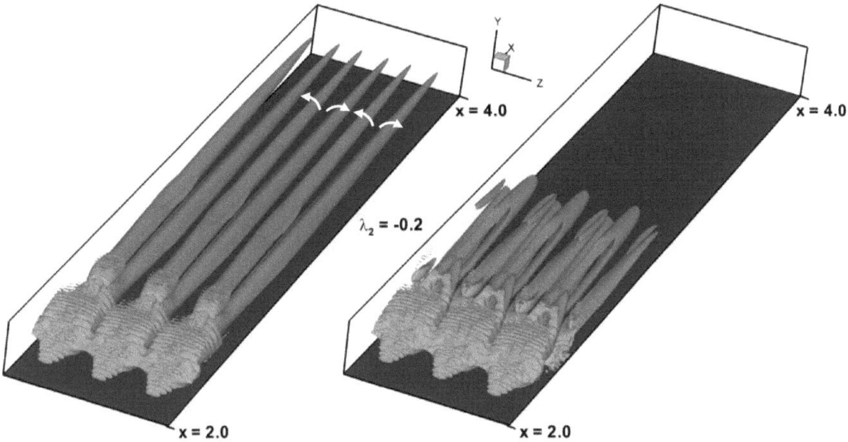

Fig. 7. Visualisation of vortical structures via λ2-isosurface (λ_2 = -0.2) for aligned rows of holes (A) and for staggered rows of holes (B). The arrows indicate the rotation sense.

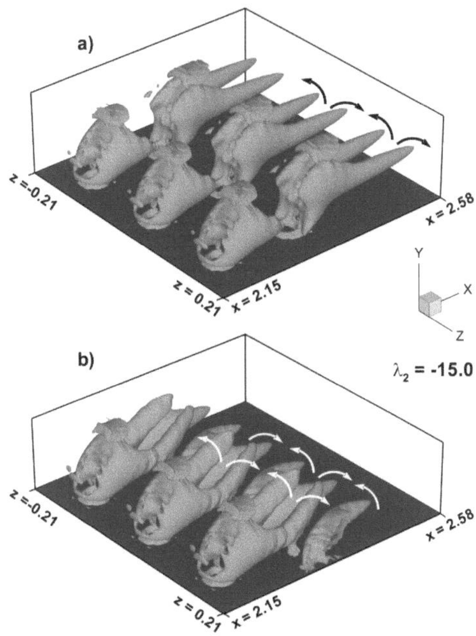

Fig. 8. Visualisation of vortical structures via λ2-isosurface (λ_2 = -15.0) for a) aligned rows of holes and b) staggered rows of holes. The arrows indicate the respective rotation sense.

transported away from the wall in the streamwise hole center line. Furthermore exists a toroidal neck vortex at each hole edge. It has a counter-clockwise rotation sense in the center-line plane upstream the hole, in contrast to the considered slit case. A horseshoe vortex is not observed in the simulations due to the low blowing ratio. In studies of jets in crossflow (JICF), where typically a horseshoe vortex (with a rotation sense opposite to the CVP and neck vortex) is found, $(\rho v)_{c,max} = O(1)$ and $d > \delta$. In the aligned case (Figs. 7a, 8a), the second row enhances the CVPs from the first row and the vortices lay wall-parallel. In contrast, Figs. 7b, 8b show the CVPs from the second row pushed downwards, keeping the cold gas at the wall. Moreover, the CVPs of both cases decay downstream.

For a Mach-6 boundary layer, the strongest amplified disturbance mode is the 2nd mode as a 2-d wave (spanwise wave number - $\gamma = 0$). We found some other almost neutral eigenvalues in the region of the slits, however the eigenvalue of the 2nd-mode disturbance is the by far most amplified in this region. Figure 9 shows the stability diagram for case A. The 1st mode (2-d) is completely stabilised because of the cooled wall and is not present anymore in the considered streamwise region. The 2nd mode is shifted to lower frequencies in the region of the slits, and the maximum amplification rate $\alpha_{i,max}$ is about twice as large as in the case without blowing. A small additional instability region develops at the slits over the 2nd mode, which seems a higher viscous mode.

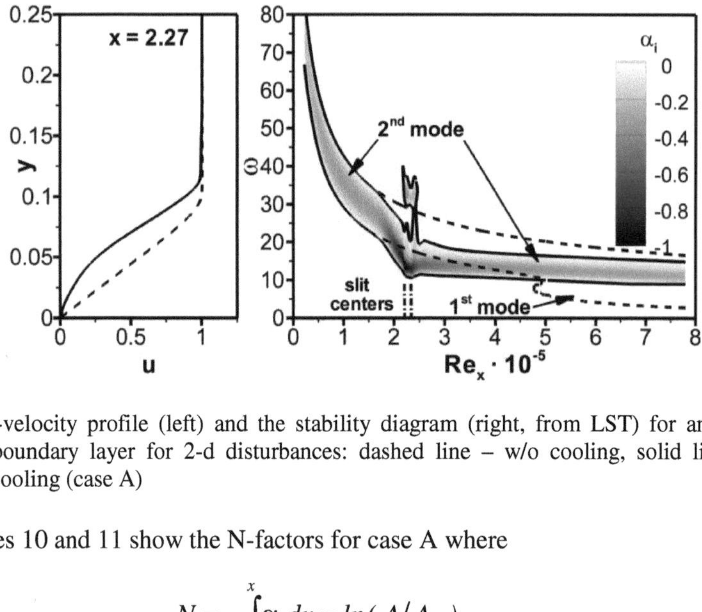

Fig. 9. u-velocity profile (left) and the stability diagram (right, from LST) for an adiabatic Mach-6 boundary layer for 2-d disturbances: dashed line – w/o cooling, solid line – with effusion cooling (case A)

Figures 10 and 11 show the N-factors for case A where

$$N = -\int_{x_0}^{x} \alpha_i dx = ln(A/A_0). \qquad (16)$$

For 2nd-mode 2-d disturbances the N-factor of the frequency $\omega = 12.5$ is approximately four times higher at the end of the considered streamwise domain than without blowing. For 1st-mode 3-d disturbances (Fig. 11) a stabilisation by effusion cooling can be seen. This is non-trivial since not only the wall but also the boundary layer itself is cooled, and, at the same time, the *u*-velocity profile has an inflection point. Primary LST uses the assumption that the spanwise base-flow gradients are

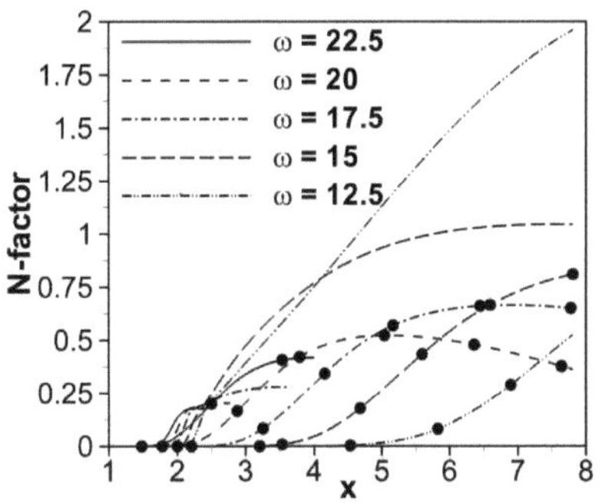

Fig. 10. N-factors for 2-d disturbances from case A with blowing (lines) and without blowing (lines with circles) for various frequencies, ω = 15 is $f^* = 74.67$kHz

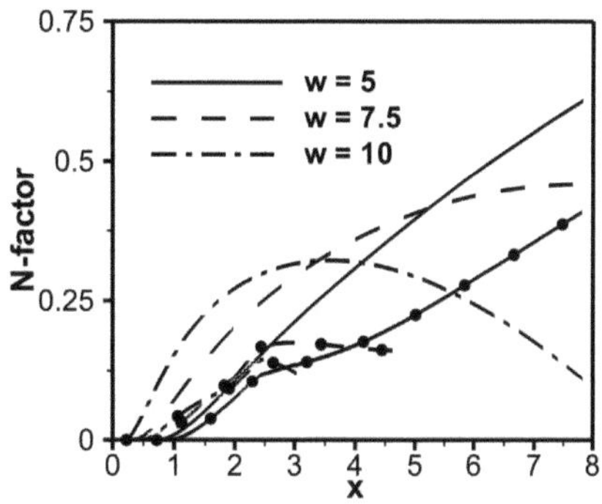

Fig. 11. N-factors for 3-d disturbances ($\psi = \arctan(\gamma/\alpha_r) \approx 60°$) from case A with blowing (lines with circles) and without blowing (lines) for various frequencies

zero. Thus it can not be used to predict the instability of cases C, D. Recall that enhanced laminar instability can compromise the cooling effect.

3.3 Instability Investigations of an Effusion-Cooled Adiabatic Mach 6 Boundary Layer

Here we investigate the same adiabatic Mach-6 boundary layer as in the section before, but now we blow cold air through four spanwise rows of holes and add unsteady 2-d background disturbances prescribed in front of the holes at the wall. Four rows of aligned holes are used, because of the stronger and persistent flow deformation of the boundary layer. Two cases are investigated: case E with a small spanwise spacing ($s_{z,a} = 0.1378 = 3d$), and case F with four times larger spacing ($s_{z,b} = 4 \cdot s_{z,a}$). The hole diameter d, the cooling gas temperature $T_{c,core}$ and the blowing ratio $(\rho v)_{c,max}$ are in both cases equal, corresponding to case B of section 3.2, Table 1. Thus the massflow through the holes in case E is only one quarter of case A per spanwise unit. The hole region lies within $2.205 \cdot 10^5 \le Re_x \le 2.756 \cdot 10^5$. A crosscut of the u-velocity field downstream the rows is shown in Fig. 12. In case F, right, the boundary layer is deformed stronger than in case E, left, both showing mushroom-like structures by the action of the CVPs.

Upstream of the holes, a packet of unsteady 2-d disturbance waves is generated by timewise periodic suction and blowing within a disturbance strip at the wall ($Re_x = 1.78 \cdot 10^5$) for a bunch of frequencies to check for laminar instability. Note that due to the large steady vortices 3-d unsteady disturbances are nonlinearly generated with the 2-d packet, and that due to the physically fixed streamwise extent of the strip, matched approximately to the streamwise wavelength of the ($\omega = 10$)-disturbance, the receptivity is lower for other frequencies. Figure 13 shows the downstream development of the u-disturbance amplitudes (u'_h – maximum over y and z) from a timewise Fourier analysis for both cases. The curve (0,0) represents the timewise and spanwise mean deformation of the 2-d boundary layer and the other curves represent the maximum over y and z of the u-disturbances (2-d and 3-d together) for a specific angular frequency $\omega = 2\pi \cdot f^* \cdot L^* / u^*_\infty$ ($\omega = 10$ is $f^* = 49.78$ kHz). In case E the mean flow deformation (0,0) is stronger than in case F due to the higher injected mass flow, but all frequencies are damped or neutral for $x > 5$, except frequencies near $\omega = 10$, being also amplified in the pure 2-d base flow, see Fig. 9. Low frequencies are neutral or damped like in case E. (we checked down to $\omega = 1$). Thus the steady 3-d flow deformation by blowing does not lead to instant transition in the young boundary layer in the front part of the plate, and a small spanwise hole spacing is preferable due to larger cooling effectiveness and lower amplification of unsteady disturbances. "Isolated" holes give rise to enhanced instability.

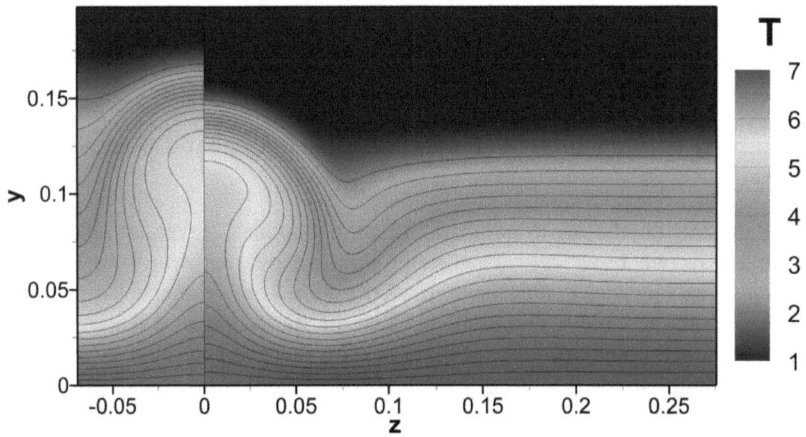

Fig. 12. Temperature field and u-velocity isolines in a crosscut at $Re_x = 3.128\ 10^5$ downstream of the holes for case E (left) and case F (right). Half the spanwise domain width is shown.

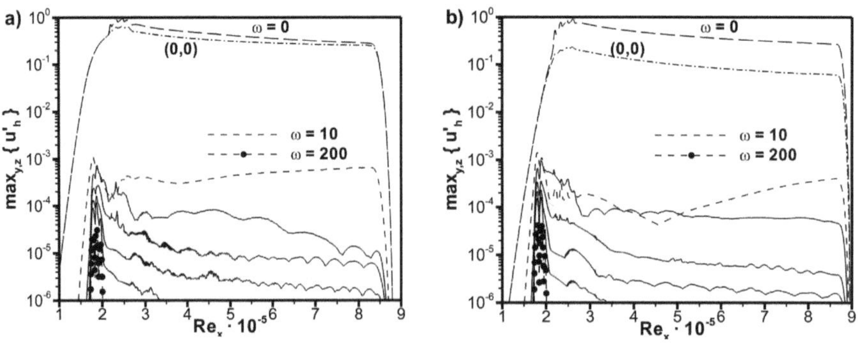

Fig. 13. Downstream t-modal amplitude development (u'_h – maximum over y and z) for cases E and F, $\omega = 10$ is $f^* = \omega u^*_\infty/(2\pi L^*) = 49.78$kHz

3.4 Effusion Cooling by Slits at Flight Conditions for a Mach 6.8 Boundary Layer and Inclined Blowing

In this section we investigate an effusion-cooled Mach 6.8 boundary layer at flight conditions at an altitude of 33 km, corresponding to a flight point of the hypersonic "Sänger" lower-stage vehicle [4]. We prescribe a radiation-adiabatic wall and $T_\infty^* = 231.45$ K, thus we have $T_{rec}^* = 2031$ K, and with $\varepsilon = 0.8$ we get $T_w \approx 0.4\, T_{rec} \approx 3.5\, T_\infty$, corresponding to 930 K at $x = 1$ decreasing to 750 K at $x = 9$ without blowing, and $p_\infty^* 0.00558$ bar, $L^* = 85.367$ mm and $Re_{unit}^* = 1.17 \cdot 10^6$ 1/m; T_c^* is again 293 K. The dimensional scales of the two slits and slit positions are the same as in section 3.2 (case A), but at flight conditions we have a different unit Reynolds number. Thus the reference length changes and the dimensionless scales are different. The geometrical data are listed in Table 2. Here also the effect of non-normal blowing is considered.

Table 2. Parameters of the 2-slits configurations for the flight-condition cases

case	$(\rho u)_{c,max}$	$(\rho v)_{c,max}$	blowing angle ϕ_c	slit width d	streamwise spacing s_x
G	-	0.0283	90°	0.0234 = 2mm	0.0585
H	0.02	0.02	45°	0.033 = 2.83mm	0.0585

In case *H* inclined blowing by an inclined duct is applied, see Fig. 14. The slit width $d_{inclined} (= d / cos\phi)$ is larger and the wall-normal mass flux $(\rho v)_{c,max} (= (\rho v)_{c,max,0°} \cdot cos\phi)$ is lower than in case A. Of course, the injected mass flow is identical, only the blowing area is enlarged and the wall-normal mass flux lowered. However we additionally have a wall-parallel flux $(\rho u)_{c,max}$ at the wall, somewhat like a locally moving wall within the slits.

The wall-temperature distribution is shown in Fig. 15. The cases with wall-normal (G) and inclined blowing (H) look similar. The reason may be the low blowing rates.

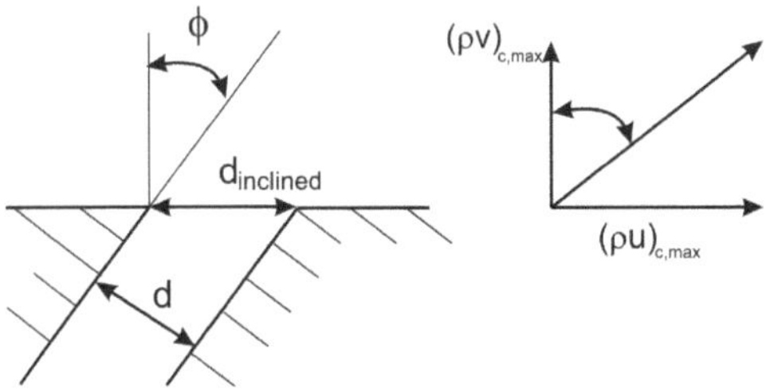

Fig. 14. Sketch of the inclined blowing

For higher blowing rates we expect the inclined blowing to yield a lower wall temperature. Note that the slit width is about $0.5 \cdot \delta$ for case G, and that $10 \cdot \delta$ downstream of the second slit ($x = 1.5$) the wall is still cooled by about 70K despite the small injected mass flow of about 3% of the total boundary-layer mass flow.

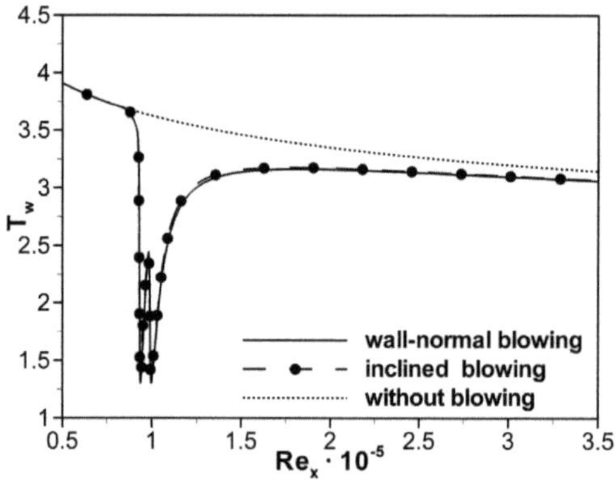

Fig. 15. Streamwise wall-temperature evolution for 2-slits-blowing (cases G and H of Table 2) for a radiation-adiabatic Mach-6.8 boundary layer

Figure 16 shows longitudinal cuts of the temperature field for both cases. At first look the temperature fields seem similar, too. Only near the slits the temperature is slightly lower in case G. The streamlines coming out of the slits show a lower angle than 45° in case H due to the stretched y-coordinate.

In Fig. 17 the downstream evolution of the cooling effectiveness η by blowing through two slits is shown for the case A at wind tunnel conditions (WTC) and for case G at flight conditions. The cooling effectiveness η is defined by

$$\text{WTC:} \quad \eta_{ad} = \frac{T^*_{rec} - T^*_{w,c}}{T^*_{rec} - T^*_{c,core}} \tag{17a}$$

and

$$\text{FC:} \quad \eta_{rad} = \frac{T^*_{w,rad} - T^*_{w,c}}{T^*_{w,rad} - T^*_{c,core}} \tag{17b}$$

where T^*_{rec} is the recovery temperature, $T^*_{w,rad}$ is the local wall temperature without blowing, and $T^*_{w,c}$ is the local wall temperature with blowing. Due to the blowing of cold air into the boundary layer, the wall temperature decreases, which results in a partial loss of the radiation cooling (Eq. (14)) at FC. Thus the cooling effectiveness in this case is lower than at wind tunnel conditions.

The amplification rate of 2nd-mode disturbances is for a radiation-adiabatic boundary layer higher than for an adiabatic boundary layer *at the same free-stream conditions*. The reason lies in the destabilization of the 2nd mode due to wall cooling.

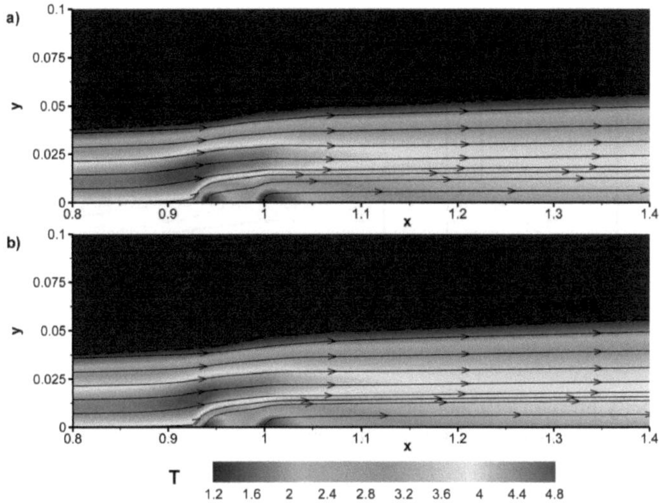

Fig. 16. Temperature fields and streamlines in a longitudinal cut for the radiation-adiabatic Mach-6.8 boundary layer with wall-normal blowing (case G) and inclined blowing (case H) - see Table 2). Dashed-dotted line: u = 0.99.

Figure 18 shows the stability diagram for case G. The amplification rates of case A can not be directly compared with the values of Fig. 9, because in both cases we use different length scales for non-dimensionalisation ($\alpha_i = \alpha_i^* \cdot L^*$). The maximal dimensional rate is $\alpha_i = -0.0293$ 1/m ($\alpha_i \cdot \delta = -0.125$) for the case at flight conditions, and for the case at wind tunnel conditions (section 3.2) $\alpha_i = -0.0276$ 1/m ($\alpha_i \cdot \delta = -0.12$). Note that the difference is not that large because of the principally "destabilizing" lower free-stream temperature in the wind-tunnel case.

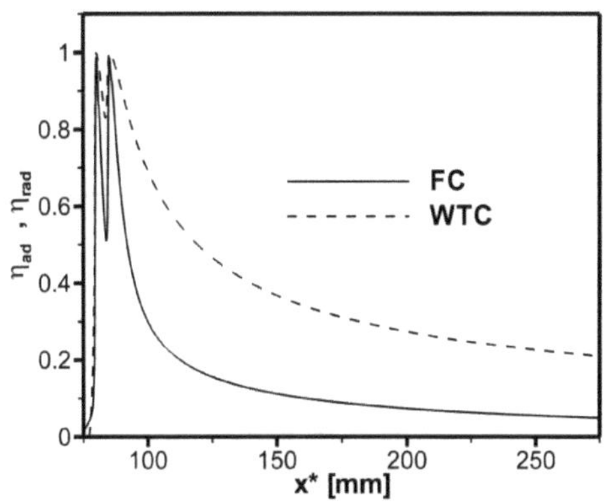

Fig. 17. Downstream evolution of the cooling effectiveness η for case A at wind tunnel conditions (WTC, adiabatic) and for case G at flight conditions (FC, radiation adiabatic)

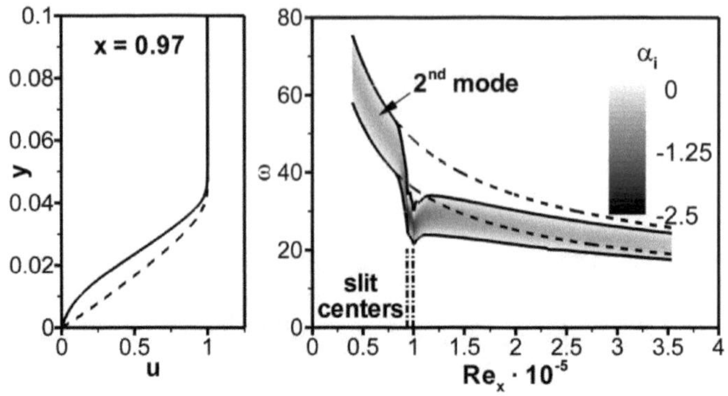

Fig. 18. u-velocity profile (left) and the stability diagram (right, from LST) for an radiation-adiabatic Mach-6.8 boundary layer for 2-d disturbances: dashed line – w/o cooling, solid line – with effusion cooling (case G)

By cool blowing the instability region is shifted to lower frequencies as in the adiabatic case in section 3.2. No higher viscous modes arise at flight conditions and no 1st mode exists in the considered streamwise region. This is due to the low Reynolds-number region we look at. Some N-factors are shown in Fig. 19 ($\omega = 25$ is $f^* = 96.7$ kHz). For case A, the maximum N-factors are two-and-a-half times higher than in the case without blowing, but are still very low due to the low local Reynolds number.

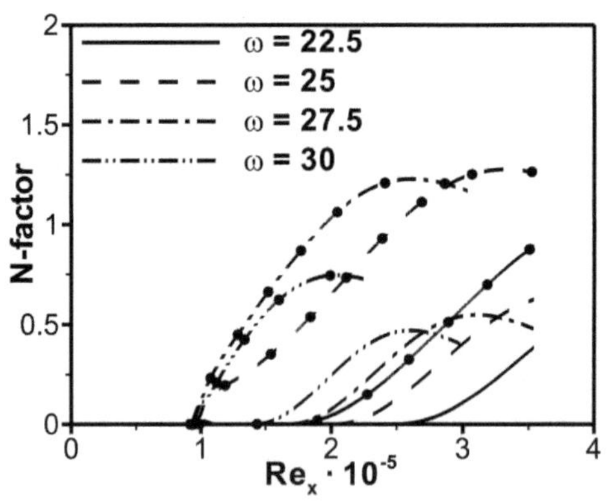

Fig. 19. N-factors for case G with (lines with dots) and without blowing (lines) for various frequencies, ω = 25 is f* = 96.7 kHz

4 Conclusions

Effusion cooling by slits and holes in various laminar zero-pressure-gradient super- and hypersonic boundary layers has been investigated using direct numerical simulations (DNS). Focus of the study using air as flow and cooling gas is not only the investigation of the cooling effectiveness of various cooling configurations for a supposedly laminar flow but also the alteration of the laminar stability properties. Enhanced laminar instability by inducing shear layers with blowing and cooling the hypersonic boundary layer can compromise the cooling if the flow transitions to turbulence due to the cooling.

The comparison between numerical and experimental results for effusion cooling of a laminar, basically already cooled isothermal Mach 2.67 boundary layer through slits shows good agreement. The simulation predicts a slightly higher cooling effectiveness for low blowing rates which may be caused by differing boundary-layer parameters that were not available from the experiment.

The presented results for effusion-cooling configurations with successive slits and holes of an adiabatic Mach 6 boundary layer at wind-tunnel conditions show that slits are better than (a few) holes. The slit-blowing velocity is, at same injected massflow, smaller than that of (a few) holes, where we had blowing ratios $(\rho v)_{c,max}/(\rho u)_\infty$ of 3% for the slits and 15% for the holes. At lower blowing ratios the coolant gas keeps closer to the wall. However our model prescribes a fixed cooling-gas temperature that does not depend on the blowing ratio. Thus wide effusion areas with low blowing are beneficial, a finding not necessarily always true for blowing in practice. Here the cooling-gas temperature will rise in time at low blowing due to the heating of the porous ceramics it has to pass. The steady state may then be different.

The analysis of the two-dimensional flow with slit blowing with primary linear stability theory shows that the maximum modal amplification rate of 2nd-mode disturbances rises by a factor of 2 and the amplified frequency band is shifted to lower frequencies for the adiabatic case. The 1st mode is completely stabilised despite the prononunced inflection point in the decelerated u-velocity profile.

Aligned rows of holes induce a strong spanwise variation of the wall temperature and less cooling effectiveness compared to staggered rows of holes. A counter-rotating longitudinal vortex pair (CVP) is generated at each hole, decaying downstream, that pushes the coolant gas of the successive, staggered row down to the wall.

For the instability analysis of the real 3-d flow field, unsteady background disturbances have been added upstream of the holes at the wall and their timewise and downstream evolution computed by DNS. In a case with four rows of aligned holes it turned out that a small spanwise spacing of the holes is preferable over a large spacing, i.e. that the spanwise spacing should roughly be less than 2.5 boundary layer thicknesses (δ) for hole diameters less than 0.5·δ ($s_z < 5d$). The steady 3-d deformation is than less detrimental. In the represented cases, no transition is observed in the young Mach 6 boundary layer with the aligned row of holes, i.e. no explosive "secondary" instability of the mushroom-like structures generated by vortices is observed. The results show only a destabilisation effect of the large spanwise spacing.

For a case at flight conditions with radiation-adiabatic wall, where the slit geometry has been adopted from the previous adiabatic case, blowing through two wall-normal and two inclined slit ducts show similar results for the flow field and the wall temperature. The reason lies in the low blowing rates. At higher blowing rates the inclined blowing will keep the coolant gas closer to the wall due to a reduced wall-normal mass-flux at enlarged blowing area, and an additional wall-parallel component. The cooling effectiveness by blowing for a radiation-adiabatic wall at flight conditions is lower then at wind tunnel conditions, due to the lower wall temperature, which decrease the radiation cooling of the wall. The stability diagram and the N-factors for the case at flight conditions show the same behaviour as in the adiabatic case at wind tunnel conditions. The N-factors are lower here because the local Reynolds number at the kept slit positions scales with the unit Reynolds number that is 2.4 times lower in this case. Transition in the considered 2-d case is unlikely due to the young boundary layer.

It appears that effusion cooling at low blowing rates of $(\rho v)_{c,max}/(\rho u)_\infty < 5-10\%$ does not significantly increase laminar instability, at least as for modal growth of disturbances. This is especially true for slits and narrow-spaced holes. If the blowing is more localized and stronger, longitudinal vortices are generated that lead to strong 3-d deformations of the mean flow.

Future investigations will focus on the effects of pressure gradients and 3-d boundary layers with crossflow. With crossflow, blowing through holes may be more dangerous as it can generate unstable crossflow vortices by modal growth. Also, effects of binary gas mixtures will be investigated.

Acknowledgements

The partial financial support of this work by the Helmholtzgemeinschaft HGF within project A8 of the RESPACE group is gratefully acknowledged. We thank the Höchstleistungsrechenzentrum Stuttgart (HLRS) for provision of supercomputing time and technical support within the project "LAMTUR".

References

1. Babucke, A., et al.: Direct numerical simulation of shear flow phenomena on parallel vector computers. In: Resch, M., et al. (eds.) High Preformance Computing on Vector Systems, Proc. High Performance Computing Center Stuttgart, pp. 229–247. Springer, Heidelberg (2003)
2. Bierbach, M.: Untersuchungen zur aktiven Kühlung der Grenzschicht-strömung an einem Plattenmodell. Diplomarbeit, Technische Universität Darmstadt (2003)
3. Eißler, W.: Numerische Untersuchungen zum laminar-turbulenten Strömungsumschlag in Überschallgrenzschichten. Dissertation, Universität Stuttgart (1998)
4. Hirschel, E.H.: Basics of Aerothermodynamics. Springer, Heidelberg (2004)
5. Jeong, J., Hussain, F.: On the identification of a vortex. J. Fluid Mech. 285, 69–94 (1995)
6. Kloker, M.: A robust high-resolution split-type compact FD-scheme for spatial direct numerical simulation of boundary-layer transition. Applied Scientific Research 59, 353–377 (1998)
7. Linn, J., Kloker, M.J.: Numerical investigations of effusion cooling in hypersonic boundary-layer flow. In: Tropea, C., et al. (eds.) Notes on Numerical Fluid Mechanics and Multidisciplinary Design, vol. 96, Springer, Heidelberg (2007)
8. Malik, M.R.: Prediction and control of transition in supersonic and hypersonic boundary layers. AIAA-J 27, 1487–1493 (1989)

9. Stemmer, C., Kloker, M.: Interference of wave trains with varying phase relations in a decelerated two-dimensional boundary layer. In: Wagner, S., Kloker, M., Rist, U. (eds.) Recent Results in Laminar-Turbulent Transition. NNFM, vol. 86, pp. 91–110. Springer, Heidelberg (2003)
10. Thumm, A.: Numerische Untersuchung zum laminar-turbulenten Strömungsumschlag in transsonischen Grenzschichtströmungen. Disseration, Universität Stuttgart (1991)
11. White, F.M.: Viscous Fluid Flow, 2nd edn. Series in Mechanical Engineering. McGraw-Hill, New York (1991)

Transpiration Cooling Methods for the SpaceLiner

A. van Foreest, M. Sippel, A. Gülhan, and B. Esser

Deutsches Zentrum für Luft- und Raumfahrt (DLR)
Systemanalyse Raumtransport (SART)
Linder Höhe, 51147 Köln, Germany
Arnold.vanForeest@dlr.de

Summary

At the Space Launcher System Analysis (SART) department of DLR-Cologne, a hypersonic spaceplane for passenger transportation is being investigated. The spaceplane is called the "SpaceLiner". The vehicle performs its rocket powered, intercontinental flight via a suborbital trajectory. The major challenge is the aerodynamic heating of the vehicle. This is discussed, and a possible solution for handling the extreme heatloads will be presented. The solution involves an innovative new way of transpiration cooling, using liquid water.

1 Introduction

1.1 The SpaceLiner

Since the end of the Concorde flights, intercontinental passenger transportation is subjected to low speed, long duration flight. However, interest in high speed supersonic and hypersonic passenger transportation is still alive. For hypersonic aircraft, the airbreathing SCRAM jet is usually seen as a promising option. Although it may be promising, practical implementation is still far from feasible. An alternative is the use of a rocket powered vehicle. The vehicle would then perform its intercontinental flight via a suborbital trajectory. Such a suborbital trajectory would have another advantage; passengers would actually fly into space. The current developments regarding space tourism indicate that this could be beneficial for the success of such a suborbital vehicle.

The SpaceLiner [4, 5, 10] is an example of such a rocket powered vehicle. The vehicle is designed to be able to fly the distance from Sydney to Western Europe, carrying 50 passengers. A picture of the SpaceLiner is given in Fig. 1. The SpaceLiner consists of two stages, a winged booster stage and a second stage, called the orbiter. The SpaceLiner is designed for vertical take off, much like the Space Shuttle does. The design was based on the requirement that the vehicle should be completely reusable. There are no solid boosters present, the booster stage and orbiter both use LH2-LOX powered engines. After separation of the two stages occurs, the booster makes a controlled re-entry and returns to the launch site. The orbiter then accelerates further to about 6.7 km/s and an altitude of 100 km. After this velocity is reached, all the fuel has been used and the remaining part of the flight is powerless.

A. Gülhan (Ed.): RESPACE – Key Tech. for Reusable Space Sys., NNFM 98, pp. 170–185, 2008.
springerlink.com

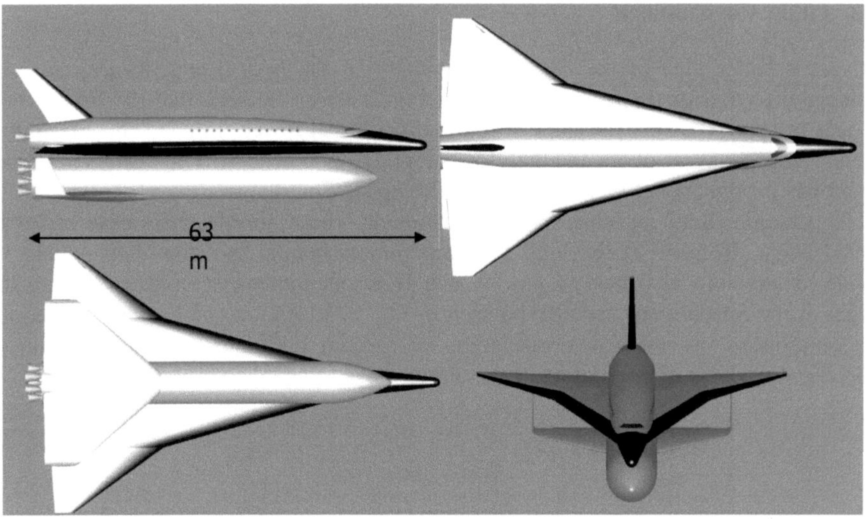

Fig. 1. SpaceLiner

By using a so called 'skip' trajectory, the range covered by powerless flight is greatly improved as compared to a ballistic trajectory. During a skip trajectory the vehicle enters the atmosphere, creates lift and leaves the atmosphere again. This is followed by a ballistic arc where after the vehicle enters the atmosphere again. The process is repeated until the trajectory converges to a gliding flight. A skip trajectory can be compared to a stone skipping over a pond, when thrown with the right speed and under the right angle. The reference trajectory flown by the SpaceLiner (Sydney-North Sea) is presented in Fig. 2. As can be seen, the vehicle begins its skip trajectory at an altitude of 100 km and with a velocity of 6.7 km/s. When an altitude of about 47 km is reached, enough lift is created to leave the atmosphere again. At this altitude, speed drops from 6.7 km/s to about 6.5 km/s, or about Mach 20. After about 3000 seconds, the skip trajectory has converged into a steady, gliding flight. After only 4500 seconds the SpaceLiner has reached its destination.

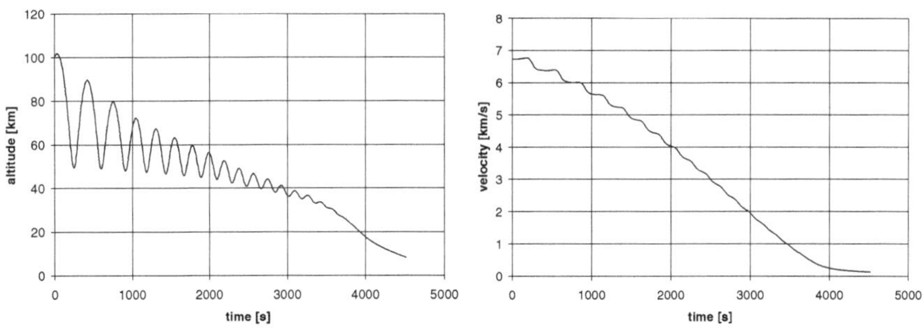

Fig. 2. Skip trajectory, altitude (left) and velocity (right)

1.2 Flight Environment

To get a better idea of the flight environment of the SpaceLiner, its trajectory is compared to that of the Space Shuttle. In Fig. 3 it can be seen that the SpaceLiner travels in approximately the same speed regime, but at lower altitude. This off course means a denser atmosphere and therefore more extreme heating. This results in higher heatloads for the SpaceLiner compared to the Space Shuttle.

Hypersonic flight introduces flow phenomena which are absent in case of lower speed flight. Because of the high air temperatures behind the shock, air cannot be modeled anymore as a perfect gas. Which flow phenomena are present during the flight of the SpaceLiner, can also be seen in Fig. 3. Vibration and excitation energies are introduced, as well as dissociation of oxygen and nitrogen. When doing a numerical analysis of the heating, these effects have to be taken into account.

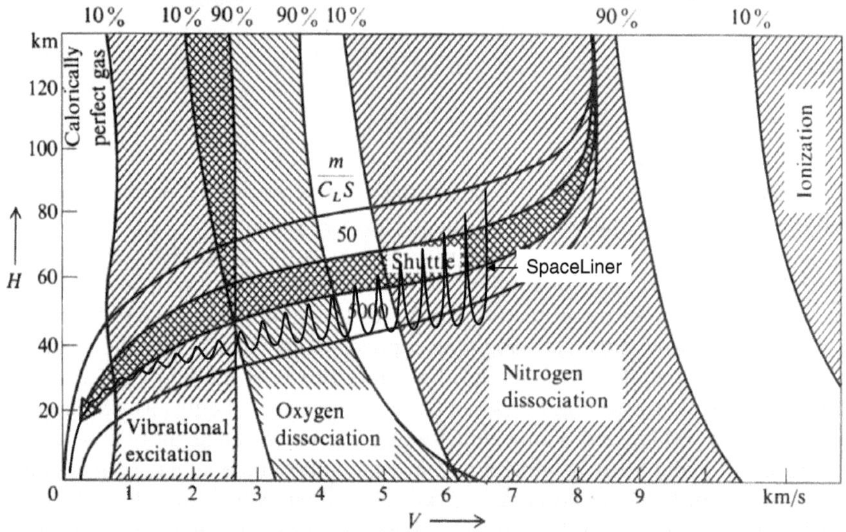

Fig. 3. Flight environment [1]

At the body surface of the vehicle, temperature will generally speaking be lower than the temperature directly behind the shock. The dissociated molecules will start to recombine. These dissociation and recombination reactions take a certain amount of time. If one assumes that the velocity of the air molecules behind the shock is low enough to allow for enough time for the reactions taking place, the equilibrium gas model can be used for numerical analysis.

In case of the SpaceLiner maximum heating is experienced at an altitude of 47 km and a Mach number of 20. Heating analysis using the equilibrium gas model results in Fig. 4. The left part of the figure assumes a laminar boundary layer, whereas the right part assumes a turbulent boundary layer. As can be seen a laminar boundary layer greatly reduces overall temperature. Temperatures on the leading edges and nose are about equal in both cases and reach about 2900 K and 2400 K, respectively. Such

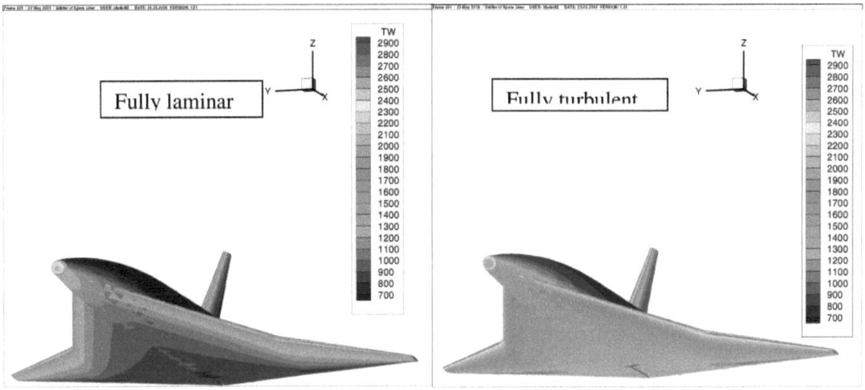

Fig. 4. SpaceLiner temperatures (K) at H = 47 km, M = 20, using the equilibrium Gas model [5]

temperatures exceed the limitations of all current thermal protection materials. Therefore, some way to reduce these temperatures has to be found.

2 Transpiration Cooling Using Liquid Water

2.1 Cooling Options

To limit the temperatures experienced by the SpaceLiner, a number of options exist. The first option is to adapt the trajectory such that heatloads decrease. Analysis shows that the initial velocity of the powerless flight phase then has to be increased to 7.5 km/s to limit heating to 1 MW/m^2 [4]. This results in a big increase in the total mass at lift off. In [4] it is stated that increase in weight would be at least 300 tons, probably even much more than this.

The second option is to change the geometry of the vehicle. For example the nose and leading edges radii could be increased. However, this would lead to a decrease in aerodynamic performance. To make up for this loss, initial speed should again be increased with the result that the weight increases by the same amount as before.

The third option is to actively cool the material down. This can be done by transpiration cooling. Transpiration cooling using a gas has been tested at DLR. To make the cooling system as light as possible, a coolant with high cooling capacity per kg has to be used. In [4, 5] it is therefore proposed to use liquid water as a coolant. Together with the wind tunnel department at DLR Cologne, a test campaign in the arc heated wind tunnel L2K has been set up to investigate the feasibility of liquid water as a coolant. In order to verify the advantage of water compared to the gas, additional tests were carried out using nitrogen gas as coolant.

2.2 Cooling Concept

Liquids will not become hotter than their boiling temperature. In case of water this boiling temperature is 100°C at 1 bar and increases proportional to the pressure. If water remains in its liquid state during the transportation through the porous material, the convective cooling will be very efficient due to the large temperature difference of liquid water and the uncooled material. When a material with a very high porosity is used, it will be cooled down to approximately the boiling temperature of the water. To prevent water from evaporating within the porous material, new water has to be supplied at a sufficiently high mass flow rate. The amount of heat, which is necessary to evaporate one kg of water, is called 'heat of vaporization'. The higher the heat of vaporization is the lower the coolant mass flow can be. Water has the highest heat of vaporization (H_{vap} = 2260 kJ/kg at 1 bar) of all liquids.

A liquid in a porous material will introduce a capillary pressure. This pressure will cause water to flow into regions where no water is present. This capillary action will therefore automatically distribute the liquid over the porous material. A simplified model of capillary action in a porous material can be made by assuming a porous material is made up of a bundle of tubes with a certain radius [9]. As soon as a capillary tube has completely filled itself with water, there will be no capillary action anymore. In case of the cooling method using liquid water, this means that when water evaporates at the surface of the material, the liquid water level in the material will drop. Capillary tubes are not completely filled with water anymore and this then causes capillary action. New water is automatically supplied to the surface at exactly the required mass flow rate.

When water evaporates at the surface, water vapour enters the boundary layer. This has an additional, advantageous effect. The water vapour 'blocks' the incoming heat flux.

A schematic view of this cooling principle is given in Fig. 5.

Before water can be evaporated, it must first heat up to the boiling temperature. This also requires some energy. This is defined by the specific heat of water, $C_{water} = 4186$ J/kgK.

Required water mass flow can be estimated without taking into account the blocking effect by using the following procedure:

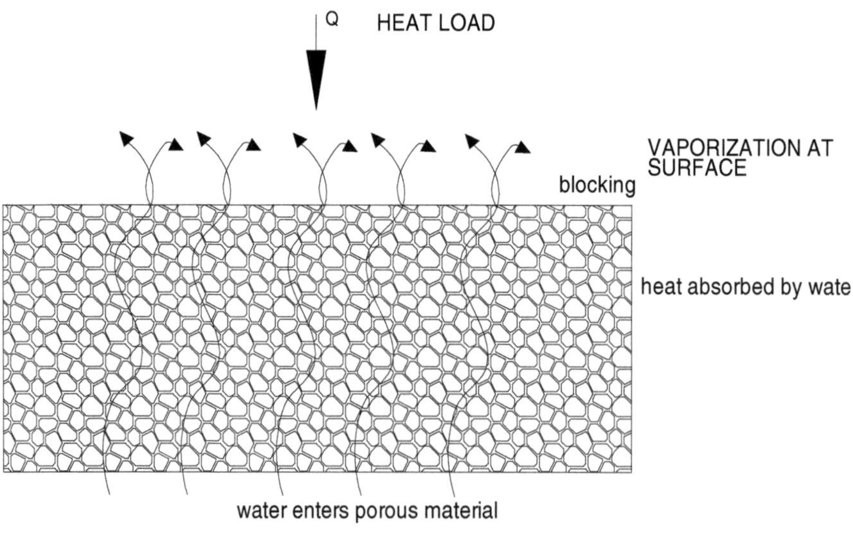

Fig. 5. Water cooling principle

Assuming the water will be supplied at a temperature of 293 K and that the boiling temperature is 373 K (at 1 bar), the temperature difference $\Delta T = 80$ K. To heat 1 kg of water up to the boiling temperature the energy supplied must be

$$C_{water} \cdot \Delta T = 334.9 \text{ kJ}.$$

As can be seen this is much less than the heat of vaporization. Therefore, for calculations only the heat of vaporization will be used for simplicity.

Water usage can then be calculated as follows

A heat balance results in the following

$$\dot{Q}_{in} = \dot{Q}_{out}$$

where $\dot{Q}_{in}$ is the heat flow [W] into the material and $\dot{Q}_{out}$ the outgoing heat flow.

Assuming all the heat is absorbed by the water we get

$$\dot{Q}_{in} = \dot{q}_{in} A = H_{vap} \cdot \dot{m} \tag{1}$$

where $\dot{q}_{in}$ is the heat flux [W/m^2], A is the heated surface area in m^2 and $\dot{m}$ the mass flow in kg/s.

Required water mass flow can then be calculated by

$$\dot{m} = \frac{\dot{Q}_{in}}{H_{vap}}. \tag{2}$$

3 Testing

The cooling concept was tested in the L2K arc heated wind tunnel at DLR-Cologne [5, 6]. Three different nose cone models were made out of a porous material called Procelit 170. This material consists of 91% Al_2O_3 and 9% SiO_2. This material was chosen because of its high porosity and its ability to withstand temperatures of up to 2000 K. The models have a varying nose radius, the smallest radius being 1 cm, the middle radius being 1.75 cm and the largest radius being 2.5 cm. The nose radius was varied to be able to investigate the influence of model geometry on the cooling efficiency. The models are shown in Fig. 6. Inside the models, a reservoir has been drilled out. The models were connected to a stagnation probe holder of L2K. A copper tube enters the reservoir for water supply. Water mass flow could be adjusted using a valve.

Tests were done using all the models. First, liquid water was used as a coolant. Temperature drops were observed for a certain water mass flow. After these tests had been completed, Nitrogen gas was used as a coolant. All the conditions were chosen identical to the other tests. The same coolant mass flow rate was used as well as the same wind tunnel flow conditions. The surface temperature was measured using an infrared camera. The test procedure was to first insert the models in the flow, without transpiration cooling switched on. Following this procedure, radiation adiabatic

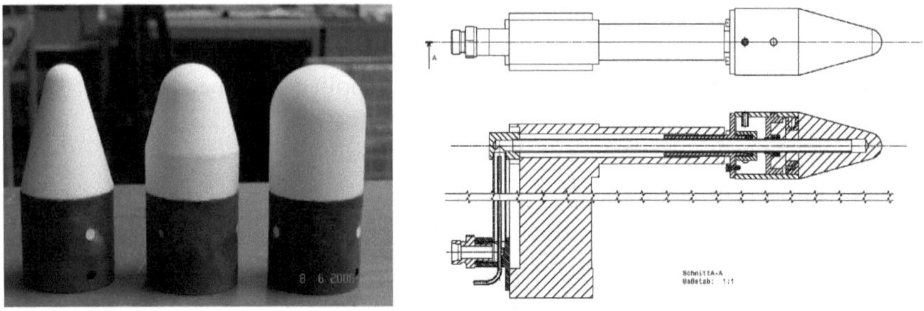

Fig. 6. Windtunnel models [5]

temperatures could thus be measured. Next, cooling was switched on and the temperature drop could be observed.

Test results of cooling using the model with nose radius of 2.5 cm are presented here. Figure 7 shows an infrared image of the temperatures in the radiation adiabatic case. As can be seen temperatures in the stagnation point reach over 2040 K. The right part of the image represents the behavior of the temperature on certain spots on the model with water cooling over time. The water mass flow rate was 0.2 g/s. Time is presented in minutes. What can be seen is that the whole model is eventually cooled to temperatures below 500 K. The infrared camera is not able to measure temperatures lower than this value, but as explained before it is expected the temperature will be equal to the boiling temperature of the water (which is about 290 K at wind tunnel conditions).

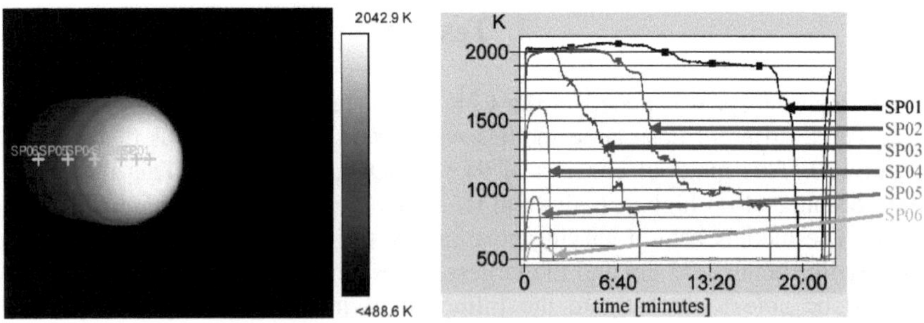

Fig. 7. Test results using 0.2 g/s liquid water [5]

The surface temperature development of the same spots using 1 g/s of Nitrogen can be seen in Fig. 8. In this case the stagnation point cooled down to about 1500 K. So even for 5 times higher gas mass flow as water, the temperature drop is still much smaller. In the right part of the figure it can be seen that for the same mass flow rate of the gas as the water (0.2 g/s), temperature drops are extremely small, especially in stagnation point regions.

Transpiration cooling using liquid water has been proven to be much more efficient compared to gas cooling. To be able to make predictions of the required water mass flow for cooling, the results have to be quantified. The first step is to determine the

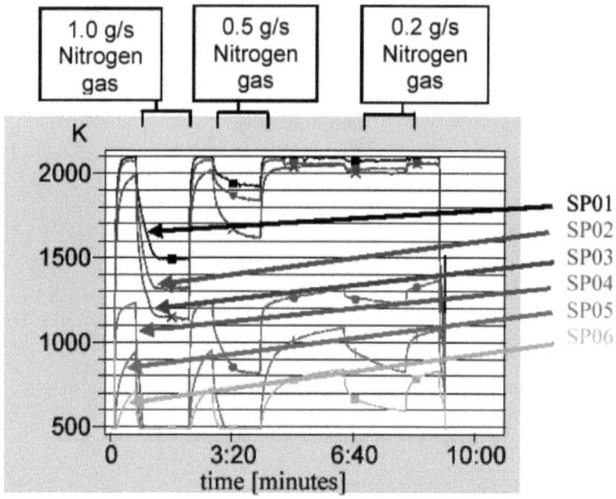

Fig. 8. Test results using nitrogen gas [5]

heat flux into the model. The heat flux then determines the evaporation rate of the water and therefore the required water mass flow. Numerical calculations for heat fluxes at wind tunnel conditions result in Fig. 9. Here the x axis represents the distance along the centerline of the model and the vertical axis represents the heat flux in W/m^2 at the surface of the model. Note that in case of radiation adiabatic conditions (cooling switched off), heat flux is much smaller than in case of a cooled wall. As explained, during the tests the model is cooled down to about 300 K. So this line is representative for the test conditions. By integrating the heat flux over the surface of the model, the total heat flow into the model can be obtained. In case of water cooling this results in 578 W. Dividing this value through the heat of vaporization of water (2460 kJ/kg at wind tunnel conditions), a required water mass flow of 0.235 g/s is calculated. This is close to the 0.2 g/s of water flow rate, which was measured during the test. The difference is due to not considering the blocking effect in calculations [5]. Further experiments and calculations showed that analysis without blocking overestimate water mass flow rate by about 30%. This then implies that even 0.2 g/s water mass flow rate is too much for this test condition.

The water cooling proved to be extremely effective. The models were cooled down from temperatures over 2000 K in the stagnation point to temperatures lower than 300 K using only little water. Compared to 'normal' transpiration cooling using a gas (in this case nitrogen), a water mass flow of only 0.2 g/s cooled the models down to much lower temperatures than was achieved for gas cooling using a coolant mass flow five times as high. In Table 1 it can be seen how extremely effective water cooling is compared to gas cooling. Here, the model with nose radius 0.025 m for water cooling and gas cooling (tests 1 and 4) are compared with each other. As can be seen, for a Nitrogen gas mass flow of 1 g/s, the stagnation point is only cooled down to 1500 K, whereas for water cooling with 0.2 g/s the stagnation point temperature drops to below 500 K. Even for 5 times as much gas as water, cooling is still less. A comparison for each measurement point on the model has been made and listed in Table 1.

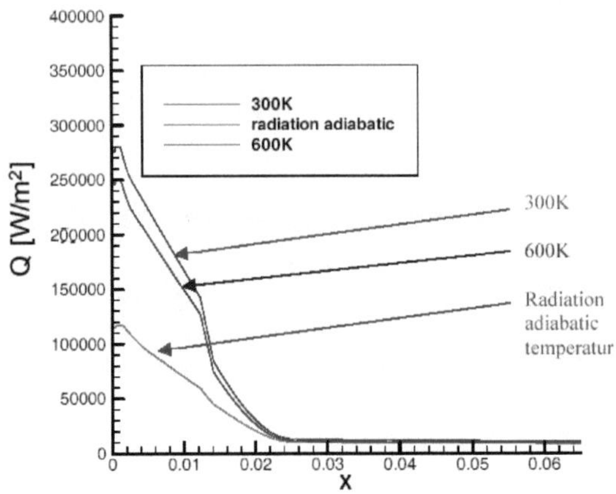

Fig. 9. Heat flux along the surface of the model [5]

Table 1. Comparison of coolants for model with R = 0.025 m

R 0.025m	Temperature drop using 0.2 g/s water	Temperature drop using 0.2 g/s nitrogen gas	Temperature drop using 0.5 g/s nitrogen gas	Temperature drop using 1 g/s nitrogen gas
Point 1	>1500 K	0 K	200 K	600 K
Point 2	>1500 K	50 K	250 K	800 K
Point 3	>1500 K	100 K	400 K	850 K
Point 4	>1100 K	100 K	400 K	>700 K
Point 5	> 450 K	300 K	>450 K	>400 K
Point 6	> 160 K	250 K	>200 K	>200 K

4 System Investigation of Water Cooling

In this chapter, the possibility and influence of application of the new cooling method on the SpaceLiner will be investigated. Using the results obtained during testing of the new cooling method, an estimation of the total water usage during the flight of the SpaceLiner could be made.

4.1 Heating Analysis

For determination of the water usage of the SpaceLiner, the heating of the SpaceLiner during its complete skip trajectory had to determined, not just the point at which the most severe heating takes place (which has been presented in section 2). The DLR trajectory analysis tool RFD calculates the heating in the stagnation point of the nose of the SpaceLiner during its complete trajectory, using a simplified formula based on the Fay-Riddell equation. For these calculations, RFD assumes a radiation adiabatic wall. To see if these results are reliable they were compared with results obtained by HOTSOSE. In order to do this, some points in the trajectory were more or less randomly picked and the heating in these points was determined by HOTSOSE, again

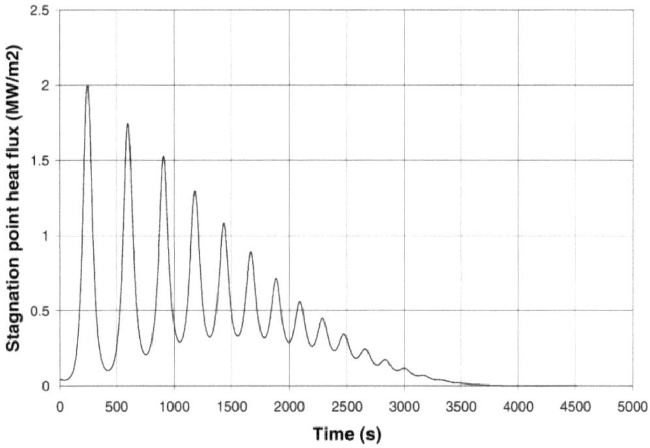

Fig. 10. Stagnation point heat flux as a function of time calculated by RFD

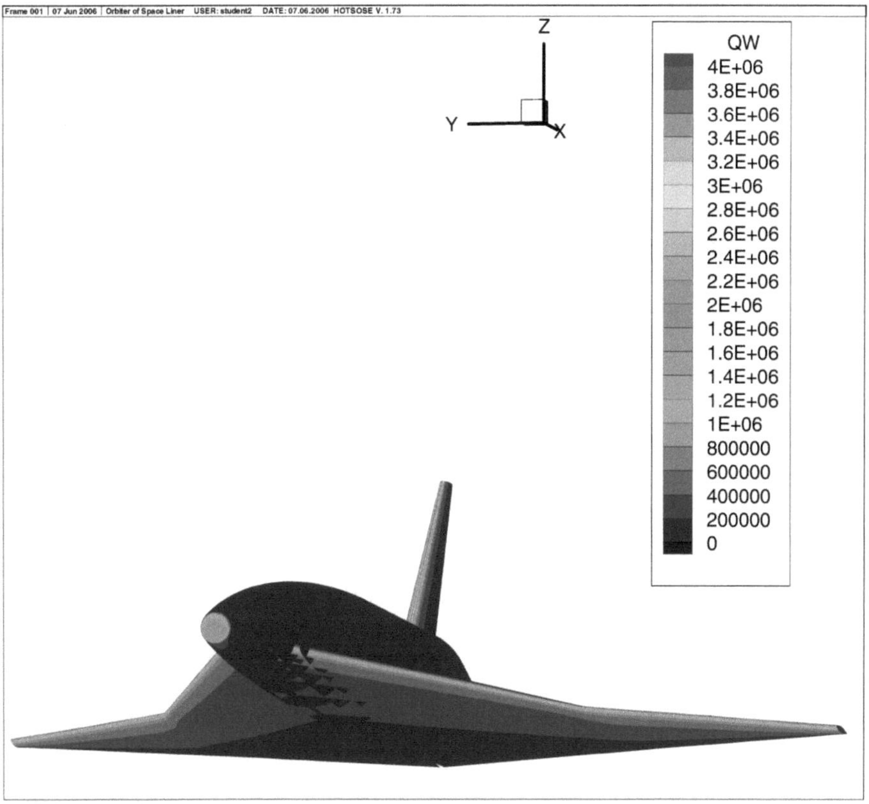

Fig. 11. HOTSOSE calculation of SpaceLiner heat flux (W/m^2) H = 47 km, M = 20

assuming a radiation adiabatic wall. The heating in the stagnation point obtained by HOTSOSE was then compared to the results of RFD, for all these different trajectory points. The results of both programs were remarkably close. This justified the use of RFD data for further analysis. Figure 10 shows the history of the stagnation point heat flux obtained by RFD for the complete trajectory of the SpaceLiner.

The next step was to determine exactly what regions of the SpaceLiner have to be cooled actively. It is assumed that the material, of which the thermal protection system is made, can withstand a temperature of 1800 K and has an emission coefficient of 0.8. This results in a heat flux of 0.48 MW/m^2. If the heat flux drops below this value no active cooling is needed. This showed that only the nose and leading edges have to be cooled actively, as can be seen in Fig. 11.

4.2 Nose Water Usage

Only when the stagnation point is subjected to heat fluxes higher than 0.48 MW/m^2, water will be needed to cool the nose.

For calculation of the total water usage, the heat flow into the complete nose of the SpaceLiner has to be determined at each point of the trajectory. According to [3], the total heat flow into a half sphere is given by

$$\dot{Q}_{total} = -\frac{4}{5}\pi R^2 \dot{q}_{sp} \cos^{\frac{5}{2}} \Theta \Big|_0^{\Theta}, 0 \leq \Theta \leq 70° . \tag{3}$$

Then, the heat flow into the nose is integrated over time, taking note of the fact that water is only needed when the stagnation point heat flux exceeds 0.48 MW/m^2. This integrated heat flow is divided by the heat of evaporation of water according to Eq. (2) resulting in the total water mass needed. Note that this does not take into account the blocking effect. Blocking will off course lead to a decrease of the total water needed, so the results obtained here can be considered conservative.

Finally, this resulted in a total of 520 kg of water needed to cool the nose.

4.3 Leading Edge Water Usage

Figure 11 shows that leading edges are also subject to heat fluxes higher than the 0.48 MW/m^2. They have to be actively cooled, too. As RFD does not calculate heating of leading edges, no time related change of leading edge heating is directly available. However, an indirect technique did allow obtaining leading edge heating.

By looking at the difference in heat flux of the nose and outboard leading edges, obtained by HOTSOSE, a difference factor could be obtained. For example, at the point of most severe heating, the heat flux into the stagnation point of the nose is about 2 MW/m^2. In the stagnation point of the outboard panel leading edge, the heat flux is about 3.4 MW/m^2. The difference factor is 3.4/2 = 1.7. Now, the values in Fig. 10 are multiplied by 1.7. The same was done for the inboard leading edge and fin leading edge, resulting in a factor of 0.7 and 1.45 respectively. This results in Fig. 12.

As in case of the nose, the heat flow into the complete leading edge has to be determined. A leading edge can be modelled as a half cylinder. Heat fluxes at points on a half circle are related to the stagnation point heat flux via

$$\frac{\dot{q}}{\dot{q}_{sp}} = \cos \Theta, 0 \leq \Theta \leq 70° .$$

Looking at Fig. 13, it can be seen that to obtain the heat flow into the complete leading edge, the following integral has to be solved

$$2\int_{0}^{70}\dot{q}(\Theta)dA\,. \tag{4}$$

With $dA = LRd\Theta$ and $\dot{q}(\Theta) = \dot{q}_{sp}\cos\Theta$, this becomes

$$2\int_{0}^{70}\dot{q}_{sp}LR\cos\Theta d\Theta = 2\dot{q}_{sp}LR\sin\Theta\Big|_{0}^{70}\,. \tag{5}$$

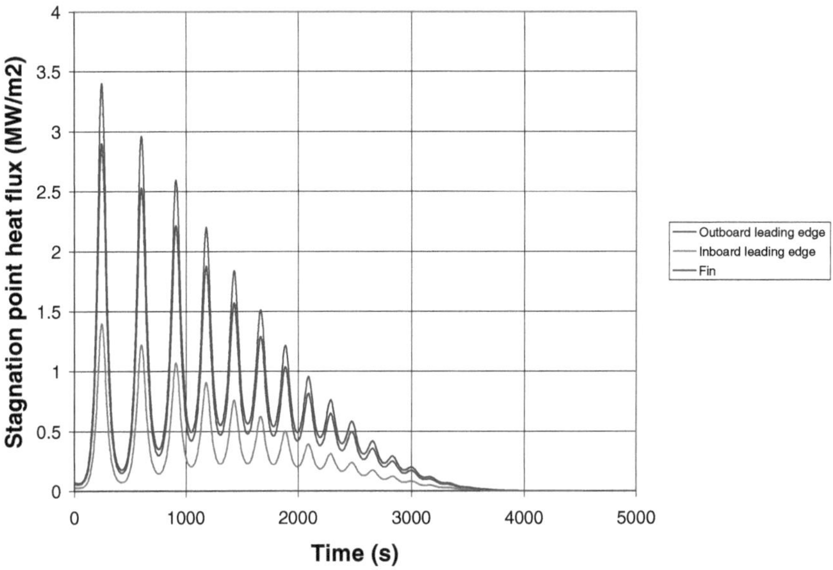

Fig. 12. Leading edge heating

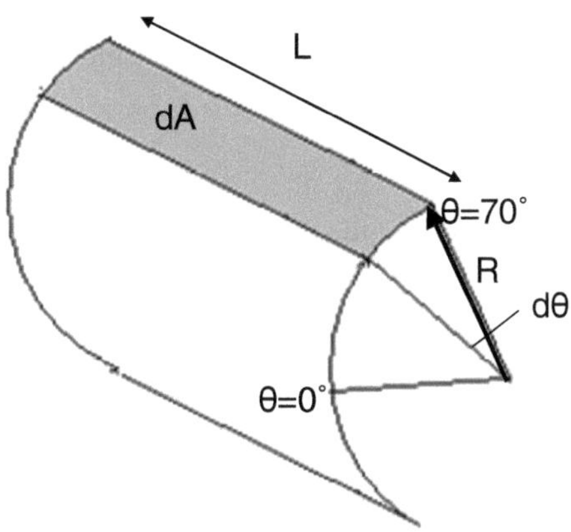

Fig. 13. Leading edge geometry

The length L of the leading edges was measured to be 17.3 m for the outboard panel, 34.5 m for the inboard panel and 12.4 m for the fin. The radius R of the leading edges varies over the span of the wing and fin. The average radius was determined to be 0.045 m for the outboard panel, 0.11 m for the inboard panel and 0.175 m for the fin.

After integrating the heat flow over time and again dividing this by the heat of vaporization this resulted in a total water usage of 2680 kg for the outboard leading edge, 3000 kg for the inboard leading edge and 2910 kg for the fin. Table 2 shows an overview of the results obtained. To cool down the SpaceLiner during its skip trajectory, some 9 tons of water is needed.

In [4] an alternative trajectory is presented which keeps the heat load on the SpaceLiner under 1 MW/m^2 without active cooling. This trajectory requires the initial velocity to be increased to 7500 m/s which leads to a mass increase of more than 300 tons. This clearly shows that active cooling using water is a very attractive option.

Table 2. Water usage

	Geometry	Radius [m]	Length [m]	Total water usage [kg]
Nose	Half sphere	0.75	-	520
Outboard leading edge	Half cylinder	0.045	17.3	2680
Inboard leading edge	Half cylinder	0.11	34.5	3000
Fin leading edge	Half cylinder	0.175	12.4	2910
Complete configuration	-	-	-	9110

4.4 Remarks on Water Usage

Cold wall heat flux and blocking effect

The calculations presented above do not take into account cold wall heat fluxes. As has been shown in the previous section, cold wall heat fluxes can be much bigger than radiation adiabatic heat fluxes.

Also, the blocking effect was not taken into account for the calculations on water usage, because it remains unclear how much the influence of this effect is on the SpaceLiner heat flux. However, what is certain is that blocking will at least partially counteract the influence of the cold wall.

Reducing the water usage

A number of methods exist to reduce the water usage of the SpaceLiner. An overview of these methods will be given in this section.

- The first method is already mentioned in section 4.3. This involves adapting the trajectory as explained in [1]. However, this option is not attractive because adapting the trajectory increases the weight of the fuel needed by far more than the water weight saved.
- The second method is to make the nose and leading edge radii smaller. This reduces the surface area that needs to be cooled and therefore reduces the water usage. However, a smaller nose radius also means an increase in heat flux and thus

an increase in water usage. Mathematically, it can be shown that the smaller area outweighs the increased heating.
According to the Fay Riddell equation [2, 3, 4, 5]:

$$\dot{q}_{sp} \mathrel{\hat{=}} \frac{1}{\sqrt{R}} .$$

Inserting this in Eq. (1) yields

$$\dot{Q}_{total} \mathrel{\hat{=}} R^{1.5} . \tag{6}$$

Now one can clearly see that the smaller the radius, the smaller the total heat flow into the nose and the smaller the water usage. In case of a nose the total heat flow and thus water usage is shown to be proportional too $R^{1.5}$.
For a leading edge the result is

$$\dot{Q}_{total} \mathrel{\hat{=}} \sqrt{R} . \tag{7}$$

In case of a leading edge the total heat flow and thus water usage is shown to be proportional too $\sqrt{R}$.

- A third method is to reduce the cold wall heat flux. This can be done by covering the porous material (such as Procelit 170) with a perforated skin. This skin would then heat up, and heat is transferred into the porous material via conduction in this skin. By choosing a material with a suitable conductivity, the skin can be allowed to heat up to a certain temperature, thus decreasing the cold wall heat flux. Perforating the skin allows water evaporated in the porous material to escape through the holes in the skin into the boundary layer, thus still being able to use the blocking effect. Figure 14 shows a schematic drawing of this concept.

  For reducing the cold wall heat flux, it is also possible to use a material with less porosity. This implies that less water can flow through the material and temperature will not decrease as much.

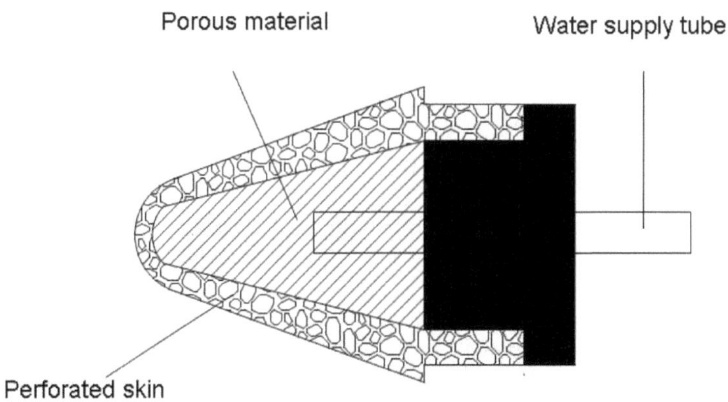

Fig. 14. Cooled nose with perforated skin

4.5 Potential Materials to Be Used

The Procelit 170 material used during the tests can not be used to make a nose or a leading edge for the SpaceLiner. The reason is that it is far too brittle. Under high stress or in case that something would hit the material (such as a bird) the material would suffer a lot of damage. A protective layer would have to be placed over the Procelit safeguarding it from damage. Such a layer can have the additional advantage of reducing the cold wall heat flux, as mentioned in the previous section. The concept showed in Fig. 14 now becomes even more interesting.

Another option is to use a porous material which is stronger and less brittle. CMC (Ceramic Matrix Composites), such as C/C and C/SiC are interesting. These materials are very strong. Temperature resistance of C/C is not very high in oxidizing atmospheres (450°C). C/SiC can withstand temperatures of up to 1750°C [10]. During manufacturing, porosity can be adapted and the required porosity can be obtained. If a material such as C/SiC is used on the SpaceLiner, porosity can be adjusted such that the material temperature is kept relatively high, thus reducing the cold wall heat flux.

5 Conclusions

To perform a flight from Sydney to Western Europe, the SpaceLiner needs to be accelerated to 6.7 km/s and an altitude of 100 km. After this acceleration phase a powerless skipping trajectory is flown. The biggest challenge of such a trajectory seems to be the aerodynamic heating. A promising new way of transpiration cooling, using liquid water as a coolant is introduced and first test results are presented. A huge increase of cooling efficiency is observed when using water instead of the option of using a gas as a coolant.

Preliminary analysis of the water usage of the SpaceLiner during its flight shows that about 9 tons is necessary to cool the vehicle down during its flight. Other options to reduce the heatload are adapting the trajectory or geometry of the vehicle. This would increase total takeoff weight by more than 300 tons. A number of ways may exist to reduce water usage, such as reducing the nose and leading edge radii and reducing the cold wall heat flux. However, more tests are needed to confirm these ideas.

References

1. Anderson, J.D.: Hypersonic and high temperature gas dynamics. McGraw-Hill Book Company, New York (1989)
2. Bannink, W.J.: Inleiding Hypersone Aerodynamica, TU Delft (1994)
3. Buursink, J.: On the Development of a Water-Cooled Metallic Thermal Protection System, TU Delft (2005)
4. van Foreest, A.: Trajectory analyses and preliminary design of a future spacecraft for intercontinental rocket powered passenger flight. DLR-SART, SART TN-008/2005 (2005)
5. van Foreest, A.: Investigation on Transpiration Cooling Methods for the SpaceLiner. DLR-IB 647-2006/05 (2006)

6. Gülhan, A., Esser, B., Koch, U.: Experimental Investigation on Local Aerothermodynamic Problems of Re-entry Vehicles in the Arc Heated Facilities LBK. AIAA J of Spacecraft & Rockets 38(2), 199–206 (2001)
7. Hirschel, E.H.: Basics of aerothermodynamics. Springer, Heidelberg (2004)
8. Reisch, U., Anseaume, Y.: Validation of the Approximate Calculation Procedure HOTSOSE for Aerodynamic and Thermal Loads in Hypersonic Flow with Existing Experimental and Numerical Results. DLR Forschungsbericht, Institut für Entwurfsaerodynamic, Braunschweig, pp. 98-123 (1998)
9. Richardson, M.: Theory and Practice in Capillary Force Vaporizer Devices. Vapore Inc. 855 C Parr Blvd Richmond CA 94801 510-235-4911 (2004)
10. Sippel, M., et al.: The SpaceLiner Concept and its Aerothermodynamic Challenges, DLR (2006)

Author Index

Notes on Numerical Fluid Mechanics and Multidisciplinary Design

Available Volumes

Volume 98: Ali Gülhan (ed.): RESPACE – Key Technologies for Reusable Space Systems - Results of a Virtual Institute Programme of the German Helmholtz-Association, 2003–2007. ISBN 978-3-540-77818-9

Volume 97: Shia-Hui Peng, Werner Haase (eds.): Advances in Hybrid RANS-LES Modelling - Papers contributed to the 2007 Symposium of Hybrid RANS-LES Methods, Corfu, Greece, 17–18 June 2007. ISBN 978-3-540-77813-4

Volume 96: C. Tropea, S. Jakirlic, H.-J. Heinemann, R. Henke, H. Hönlinger (eds.): New Results in Numerical and Experimental Fluid Mechanics VI - Contributions to the 15th STAB/DGLR Symposium Darmstadt, Germany, 2006. ISBN 978-3-540-74458-0

Volume 95: R. King (ed.): Active Flow Control - Papers contributed to the Conference "Active Flow Control 2006", Berlin, Germany, September 27 to 29, 2006. ISBN 978-3-540-71438-5

Volume 94: W. Haase, B. Aupoix, U. Bunge, D. Schwamborn (eds.): FLOMANIA - A European Initiative on Flow Physics Modelling - Results of the European-Union funded project 2002 - 2004. ISBN 978-3-540-28786-5

Volume 93: Yu. Shokin, M. Resch, N. Danaev, M. Orunkhanov, N. Shokina (eds.): Advances in High Performance Computing and Computational Sciences - The Ith Khazakh-German Advanced Research Workshop, Almaty, Kazakhstan, September 25 to October 1, 2005. ISBN 978-3-540-33864-2

Volume 92: H.J. Rath, C. Holze, H.-J. Heinemann, R. Henke, H. Hönlinger (eds.): New Results in Numerical and Experimental Fluid Mechanics V - Contributions to the 14th STAB/DGLR Symposium Bremen, Germany 2004. ISBN 978-3-540-33286-2

Volume 91: E. Krause, Yu. Shokin, M. Resch, N. Shokina (eds.): Computational Science and High Performance Computing II - The 2nd Russian-German Advanced Research Workshop, Stuttgart, Germany, March 14 to 16, 2005. ISBN 978-3-540-31767-8

Volume 87: Ch. Breitsamter, B. Laschka, H.-J. Heinemann, R. Hilbig (eds.): New Results in Numerical and Experimental Fluid Mechanics IV. ISBN 978-3-540-20258-5

Volume 86: S. Wagner, M. Kloker, U. Rist (eds.): Recent Results in Laminar-Turbulent Transition - Selected numerical and experimental contributions from the DFG priority programme 'Transition' in Germany. ISBN 978-3-540-40490-3

Volume 85: N.G. Barton, J. Periaux (eds.): Coupling of Fluids, Structures and Waves in Aeronautics - Proceedings of a French-Australian Workshop in Melbourne, Australia 3-6 December 2001. ISBN 978-3-540-40222-0

Volume 83: L. Davidson, D. Cokljat, J. Fröhlich, M.A. Leschziner, C. Mellen, W. Rodi (eds.): LESFOIL: Large Eddy Simulation of Flow around a High Lift Airfoil - Results of the Project LESFOIL supported by the European Union 1998 - 2001. ISBN 978-3-540-00533-9

Volume 82: E.H. Hirschel (ed.): Numerical Flow Simulation III - CNRS-DFG Collaborative Research Programme, Results 2000-2002. ISBN 978-3-540-44130-4

Volume 81: W. Haase, V. Selmin, B. Winzell (eds.): Progress in Computational Flow Structure Interaction - Results of the Project UNSI, supported by the European Union 1998-2000. ISBN 978-3-540-43902-8

Volume 80: E. Stanewsky, J. Delery, J. Fulker, P. de Matteis (eds.): Drag Reduction by Shock and Boundary Layer Control - Results of the Project EUROSHOCK II, supported by the European Union 1996-1999. ISBN 978-3-540-43317-0

Volume 79: B. Schulte-Werning, R. Gregoire, A. Malfatti, G. Matschke (eds.): TRANSAERO - A European Initiative on Transient Aerodynamics for Railway System Optimisation. ISBN 978-3-540-43316-3

Volume 78: M. Hafez, K. Morinishi, J. Periaux (eds.): Computational Fluid Dynamics for the 21st Century. Proceedings of a Symposium Honoring Prof. Satofuka on the Occasion of his 60th Birthday, Kyoto, Japan, 15-17 July 2000. ISBN 978-3-540-42053-8

Volume 77: S. Wagner, U. Rist, H.-J. Heinemann, R. Hilbig (eds.): New Results in Numerical and Experimental Fluid Mechanics III. Contributions to the 12th STAB/DGLR Symposium, Stuttgart, Germany 2000. ISBN 978-3-540-42696-7

Volume 76: P. Thiede (ed.): Aerodynamic Drag Reduction Technologies. Proceedings of the CEAS/DragNet European Drag Reduction Conference, 19-21 June 2000, Potsdam, Germany. ISBN 978-3-540-41911-2

Volume 75: E.H. Hirschel (ed.): Numerical Flow Simulation II. CNRS-DFG Collaborative Research Programme, Results 1998-2000. ISBN 978-3-540-41608-1

Volume 66: E.H. Hirschel (ed.): Numerical Flow Simulation I. CNRS-DFG Collaborative Research Programme. Results 1996-1998. ISBN 978-3-540-41540-4

MIX
Papier aus verantwortungsvollen Quellen
Paper from responsible sources
FSC® C105338

If you have any concerns about our products,
you can contact us on
ProductSafety@springernature.com

In case Publisher is established outside the EU,
the EU authorized representative is:
Springer Nature Customer Service Center GmbH
Europaplatz 3, 69115 Heidelberg, Germany

Printed by Libri Plureos GmbH
in Hamburg, Germany